When Worlds Quake

When Worlds Quake

The Quest to Understand the Interior of Earth and Beyond

HRVOJE TKALČIĆ

PRINCETON UNIVERSITY PRESS
PRINCETON & OXFORD

Published by Princeton University Press
41 William Street, Princeton, New Jersey 08540
99 Banbury Road, Oxford OX2 6JX

press.princeton.edu

GPSR Authorized Representative: Easy Access System Europe - Mustamäe tee 50, 10621 Tallinn, Estonia, gpsr.requests@easproject.com

All Rights Reserved

ISBN: 9780691271477
ISBN (e-book): 9780691271538

Library of Congress Control Number: 2025945775

British Library Cataloging-in-Publication Data is available

Editorial: Ingrid Gnerlich and Whitney Rauenhorst
Production Editorial: Jaden Young
Jacket/Cover Design: Karl Spurzem
Production: Danielle Amatucci
Publicity: Matthew Taylor and Kate Farquhar-Thomson
Copy Editor: Lucinda Treadwell

Jacket images: Courtesy of Solar System Scope, iStock, and Texturelabs.

This book has been composed in Arno

Printed in the United States of America

10 9 8 7 6 5 4 3 2 1

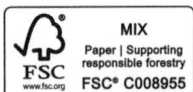

CONTENTS

ACKNOWLEDGMENTS

As I reflect on the journey that led to the creation of *When Worlds Quake*, I am profoundly grateful to the many individuals who have shaped my path in significant and subtle ways. My early years were filled with wonder, thanks to my parents, family members, and teachers, who nurtured my curiosity, and my childhood friends, who inspired my imagination through our adventures together. My academic mentors and colleagues from various places where I have lived have played a pivotal role in igniting my passion for science and, more specifically, geophysics and the mysteries of the Earth, drawing from diverse perspectives across Europe, North America, Asia, Australia, and beyond. Their insights and guidance laid the foundation for my academic pursuits and inspired me to seek answers to questions I had yet to formulate.

I extend my heartfelt thanks to the University of Zagreb, where my journey began, and to the University of California at Berkeley, which broadened my horizons and deepened my understanding of geophysics. At both institutions, I was fortunate to learn from extraordinary faculty and scientists who encouraged me to think critically and creatively. My current home at the Australian National University has further nurtured my growth, providing a vibrant research environment where innovation flourishes.

Within this supportive framework, I have been able to explore the realms of Seismology and Mathematical Geophysics, engaging in fundamental research that often transcends the confines of traditional funding projects. It is within these spaces of intellectual freedom that the most remarkable discoveries emerge—those moments of insight that not only advance science but also feed my passion for conveying our findings to the general public. The encouragement to take risks in my research has been invaluable, allowing me to pursue ideas that may not have initially seemed viable but ultimately led to meaningful contributions to our understanding of natural phenomena through observational seismology.

I want to extend my heartfelt appreciation to my research group, including my PhD students, postdocs, visitors, and all collaborators who have worked alongside me over the years. Your dedication, creativity, and enthusiasm have enriched our projects and created an inspiring environment for exploration and discovery. There is no greater reward in academia than witnessing your growth into mature researchers, each of you spreading your wings and making your marks in the field. Thank you for your hard work, passion, and the many moments of shared insight that have shaped our collective journey.

I would also like to acknowledge the dedicated members of the Australian National University Media Team. Your efforts in disseminating our geophysics discoveries to a global audience have been significant in bridging the gap between scientific research and public understanding. Your commitment to sharing knowledge ensures that our work resonates beyond the academic sphere and fosters a greater appreciation for the science that underpins our world. Special thanks to all journalists for your interest in my research, helping shape my responses, and helping me realize what is relevant to the general audience.

My Facebook profile is a place where I step out of my comfort zone by sharing knowledge and perspectives about earthquakes and geophysics in general. It could be said that the feedback I receive about my posts is a driving force that pushes me and motivates me to write more. On the other hand, perhaps these posts inspire a student or two who is still thinking about their future profession or life calling. Therefore, I want to sincerely thank each one of you who reacted and provided a response and encouragement because you contributed to this book in a smaller or more significant way.

A special acknowledgment goes to B.L.N. Kennett, whose careful reading of the manuscript draft and invaluable suggestions helped refine my thoughts and sharpen my insights. Your knowledge and encouragement have been instrumental in bringing clarity to my ideas, and I am truly grateful for your generosity in sharing your time and insights.

I also sincerely thank Ms. Ingrid Gnerlich, the Publisher for the Sciences at Princeton University Press. Your warm welcome and steadfast support have made the publishing process a rewarding experience. Your belief in this project has fueled my determination to communicate the wonders of earthquakes to a broader audience. I extend my thanks to the assistant editor, Whitney Rauenhorst, who skillfully and smoothly handled the manuscript.

Special thanks go to the copy editor, Lucinda Treadwell, for her studious and considerate copyediting and to the entire Princeton University Press production team for their exceptional work, led by Jaden Young. Their skillful coordination and dedication ensured smooth sailing for the book's publication.

Additionally, I am thankful to my Croatian publisher, Naklada Ljevak, particularly to Ms. Ivana Ljevak Lebeda, whose flexibility with rights has allowed me to navigate this journey easily. Your understanding and collaboration have made it possible for this work to reach readers far and wide, bridging the gap between cultures and scientific communities. Bonislav Kamenjašević's linguistic insights often went beyond just the language and inspired me to improve the manuscript further.

Last but certainly not least, my deepest gratitude goes to my family. Your unwavering support, encouragement, and love have been my anchor throughout this endeavor. From late-night discussions about my ideas to our shared meals and lively conversations and laughs over dinner and during our travels together, you have always been there, providing the foundation upon which I draw inspiration to write and do research. Each moment spent together, whether exploring new places or diving into the nuances of my research, has enriched my journey. It is with you in mind that I write, hoping to share the wonder of our planet and beyond with others.

Each of you has contributed to this book in direct and indirect ways. As I put these words to paper, I am reminded that this journey in pursuit of and celebrating science is not just my own; it is a tapestry woven from the threads of shared experiences, insights, and the unwavering belief in the power of knowledge and wonder. Thank you all for being part of this adventure and inspiring me to share the stories of our incredible field of science.

When Worlds Quake

Introduction

"Potres." One of my earliest childhood memories is that of my mother uttering this word in disbelief. Quake. I recall the sudden awakening, my mother's comforting embrace, and the prolonged clinking of the shaking glasses and porcelain in the living room display cabinet on the second floor of our three-story building.

My childhood was peaceful, spent in the small provincial town of Vinkovci in Croatia, nestled in the southern part of the Pannonian Plain. Known as one of Europe's oldest continuously inhabited settlements, with a history stretching back 8,000 years, Vinkovci seemed an unlikely backdrop for anything more dramatic than snow flurries, let alone something as astonishing as an earthquake. At the time, my understanding of earthquakes came from fleeting images of devastation and tragedy occurring in distant corners of the globe, pictures which occasionally flickered across our bulky television screen. In those days, such events seemed worlds away, mere curiosities far removed from the quiet rhythms of my everyday life. But that evening, it was indeed a quake that disrupted the calm.

A deep, ominous rumble, followed by an intense vibration, snapped me out of my thoughts. My tiny, wooden house in

Berkeley, California, perched precariously close to the Hayward Fault, shuddered so violently that I felt goosebumps and an uncomfortable thrill. I was experiencing another earthquake, feeling its waves ripple through me. A fascination with seismology, the science dedicated to understanding earthquakes and the waves they generate, had led me from my comparatively quiet birthplace in Europe to pursue a PhD and postdoctoral research on the west coast of the American continent, where the Earth's movements are more than just distant or rarified curiosities— they are a tangible and ever-present reality. After an initial sense of shock, I was filled with an inexplicable joy.

My ongoing research into earthquakes eventually led me yet another ocean away—to a small, top-floor flat in a Yokohama suburb, where I worked on a book I had started writing on the Earth's inner core. A small table squeezed between the bed and the wall had space for little more than my laptop, a notebook, and a pen. As evening descended, my thoughts drifted beyond the blooming cherry trees, across the harbor, and back over the vast expanse of the Pacific to where I had spent my doctoral days in the San Francisco Bay Area. Almost as though the Earth was responding to my thoughts, I felt a local quake stir, its tremors shaking the flat. It felt like some slumbering giant awakening briefly, pulling my thoughts back to the earth beneath my feet.

If not for my fascination with such stirring giants and the powerful waves they send through the planet's interior, I never would have traveled so far around the world—lastly, to the smallest of all continents, Australia. Here in Canberra, I work as a seismologist, studying quakes to better understand the inner workings of the Earth and other planets. In March 2020, my sister texted me from her home in Zagreb, the city of my student days, where a quake measuring 5.5 of Richter magnitude had struck. Like many residents of Zagreb and its surroundings, she was deeply affected by the quake, and the days that followed were filled with unease and anxiety—compounded by the fact that the city was in lockdown due to the coronavirus pandemic. From halfway around the world, I answered her—and then went on to engage in many online conversations about earthquakes with residents of Croatia, who were seeking to make sense of their experiences. When another, even more powerful earthquake struck late that same year near the town of Petrinja, not far from Zagreb, my writing began to evolve, and I decided to write a popular-level book. I felt a duty to clarify the phenomenon of earthquakes and explain the sensations caused by passing seismic waves in a deeper and more expansive way than short social media posts allow.

I originally published the book you are now reading in my native language of Croatia, but the global perspective I have gained through my travels eventually compelled me to translate the book into English in order to share a scientific understanding of earthquakes and the Earth's interior with more people. Inspired by my childhood memory of the quake that my mother and I experienced, as well as the experience of explaining earthquakes to Croatian people on social media after my sister texted me from Zagreb, I decided to use the two earthquakes that occurred in 2020 in Croatia as examples later in this book. While earthquakes are far more infrequent in the tiny European nation where I grew up than in places like Japan or California, the very nature of global seismology transcends the borders of any one country, and these two earthquakes provided an opportunity for me to explain both the personal sensation of an earthquake and some of its fundamental characteristics. I have made slight adaptations for readers who may not be familiar with Croatia and its surroundings.

Overall, the book is designed as a standalone read that, I hope, will captivate anyone curious about the science of earthquakes and the mysteries of the Earth's interior, as well as people who have experienced personal encounters with earthquakes. The book is also suitable for high school students or university students in the early stages of studying seismology. Organized into ten chapters, the book explores the history of seismology and the nature of earthquakes but also shares more personal reflections and experiences from my life as an observational seismologist. My work has brought me to some fascinating (and challenging) places—and so our quest to reveal the secrets of the Earth takes us from Eurasia and North America to the remote corners of the Australian continent, down into the icy waters of the Southern Ocean, and even beyond, to the desolate landscapes of the Moon and Mars.

In chapter 1, I place Earth within the broader context of the Solar System and the universe, examining it as a planet whose surface activity directly results from its internal dynamics. Here, we step into the realm of planetary physical processes: the formation of planets, the stratification of their interiors, the crystallization of the inner core, the phenomenon of the geodynamo, the generation of the geomagnetic field, mantle convection, plate tectonics, and the earthquakes and volcanoes that arise as a consequence of our planet's inner workings. Earthquakes are not merely disruptive events; they are witnesses to the planet's evolution and internal dynamics, messengers from the depths of time reminding us of the ceaseless activity beneath our feet. They are, in a sense, travelers through time, carrying with them the story of our planet's ongoing journey.

Chapter 2 delves into the rich tapestry of Earth science history, uncovering fascinating details about our evolving understanding of earthquakes. It begins with an ancient tale, the

Japanese legend of Namazu, the giant catfish believed to cause seismic disturbances. The narrative then moves to the late 19th century, when John Milne's invention of the first seismograph marked a significant milestone in earthquake study. The chapter climaxes with the San Francisco Earthquake and Fire of 1906, a devastating event that shifted our perception of earthquakes. It was this disaster that unveiled the true nature of earthquakes as originating from fault zones where rocks are poised to break under accumulated tension. Amid the devastation, an unusual footnote to history was written: a cow named Matilda, tragically caught in the quake, fell into a fissure along the San Andreas Fault. Her predicament made headlines, symbolizing the unexpected consequences of the event. Before this, it was widely believed that faults were the result of earthquakes rather than their cause. The San Francisco Earthquake and Fire not only redefined our understanding of earthquake mechanics but also revolutionized our view of Earth as a dynamic, active planet. This seismic event gave impetus to the rapid growth of a new geophysical discipline—seismology, a burgeoning discipline at the time—and laid the groundwork for the hypothesis and eventual theory of plate tectonics, forever altering our comprehension of the Earth's internal processes and its ever evolving nature.

Chapter 3 brings to light the monumental discoveries of the pioneers who reshaped our understanding of the Earth's interior: Andrija Mohorovičić and Inge Lehmann. While Milne laid the groundwork for modern seismology by building a seismometer, Mohorovičić and Lehmann further illuminated the Earth's hidden depths. Andrija Mohorovičić, a seismologist and meteorologist, made groundbreaking strides with his discovery of the Mohorovičić Discontinuity, or "Moho," which revealed the boundary between the Earth's crust and the underlying mantle. This revelation was a pivotal moment, offering new insights into the Earth's structure and laying the foundation for future research. A little later, Inge Lehmann advanced our knowledge even further by identifying the existence of the Earth's inner core, a discovery that transformed our understanding of the planet's internal workings. Her work built upon the discoveries of her predecessors, providing a clearer picture of the Earth's layered structure. These trailblazers did not look to the skies but instead turned their gaze downward, sliding along the verticals into the depths of our planet. Their discoveries have inspired countless Earth scientists, including myself and my colleagues, who stand on the shoulders of these giants. Their monumental contributions continue to influence and guide the field of seismology, shaping our understanding of the Earth's interior and the dynamic forces that drive its processes.

The central narratives of chapters 4 and 5 revolve around the seismic giants that stirred continental Croatia and surrounding countries in 2020—the earthquakes near the city of Zagreb and the neighboring town of Petrinja. Though seemingly unrelated, these two events occurred within the same calendar year and shook the entirety of central Europe. They serve as examples for exploring how seismologists analyze ground motion records—that is, by using seismograms to pinpoint the location and magnitude of earthquakes. The chapters delve into the scenario of these earthquakes, providing a comprehensive look at how seismologists use their tools to decode the quakes.

We journey from the earthquake's epicenter, on the crests and troughs of these waves, through the intricate workings of the Earth's layers, to the sensitive seismographs that measure ground movement. Eyewitness accounts shared on social media offer palpable illustrations of the sensations experienced in different locations and times, capturing the sound and senses of ground motion during and after the quakes. Through these accounts, we explore fundamental concepts such as the differences in earthquake magnitude, the frequency of aftershocks, the triangulation method, and the use of spectrograms and geodetic interferograms. We also discuss the nature of sound and infrasound during earthquakes and whether and how earthquakes may be temporally and spatially connected. These chapters lay the foundation for a deeper exploration of earthquake forecasting and possible prediction discussed in subsequent book sections.

From the warmth of southeast Europe, we journey to a cold, star-studded winter night over the Barents Sea, where the aurora borealis dances in shimmering waves across the upper reaches of the atmosphere. Beneath the frozen expanse of the Kola Peninsula, we find ourselves at the site of humanity's deepest borehole—a monumental feat in our quest to probe the Earth's interior. Yet, despite the impressive depth we have achieved compared with the deepest mines, it becomes clear that we have barely "scratched the surface" in our journey toward the Earth's center.

In chapter 6, we linger a bit longer with the theme of earthquakes but with a new focus. Here, modern techniques of seismogram analysis come to the forefront, revealing how we use these seismic waves to "image" the Earth's internal structure. Although earthquakes are not the primary subject of study here, they serve as the source of waves that traverse the Earth's surface and penetrate its depths. When direct exploration is constrained by the extreme temperatures and pressures of the Earth's crust and mantle, we rely on seismic waves to guide us. This chapter introduces seismic tomography, drawing parallels with medical tomography to explain how it maps the Earth's internal structure. I describe the insights we have gained about the Earth's interior, though the image remains somewhat blurry. However, our picture is becoming increasingly sharper thanks to recent and ongoing advancements, offering us a more defined glimpse into the mysteries beneath our feet.

Chapter 7 ventures into the heart of one of science's most tantalizing questions: can we predict earthquakes, and is there hope for accurate predictions in the future? The chapter opens with an account of the 1976 Tangshan earthquake—a disaster that left an indelible mark on the Chinese city and became one of the most devastating seismic events in recent history. This heart-wrenching story sets the stage for a broader exploration into the world of earthquake prediction. Our journey then takes us deeper into China's seismic history, where I examine other significant earthquakes and early prediction attempts. The tale of the Haicheng earthquake stands out, where a rare success in forecasting ignited excitement and methodical approaches to predicting quakes. Yet, the optimism that once surrounded earthquake prediction was met with harsh reality. As the devastating impacts of Tangshan taught us, predicting earthquakes proved far more complex than anticipated, leading us to reassess our methods.

With a renewed focus on scientific methods, the chapter shifts to Parkfield, California. Here, I reflect on how the pursuit of earthquake prediction has evolved, marking a return to understanding fault processes and seismic behavior. I conclude by examining modern research directions—a metaphorical "crystal ball" of contemporary science. Through this lens, I explore the innovative strides being made and the ongoing quest to unlock earthquake prediction. As we look to the future, the hope of accurately predicting earthquakes remains a compelling and elusive challenge, driving the relentless pursuit of scientific discovery.

In chapter 8, we are immersed in the fiery expanse of the Australian Outback, in the rich redness of its ancient rocks, the mystique of Aboriginal heritage, and exotic wildlife. As we traverse Stuart Highway and other remote parts of the continent, we find ourselves in a natural laboratory unlike any other. Amid this rugged landscape, we delve into the intriguing world of nuclear testing and the global efforts to detect and prevent clandestine nuclear experiments. I will take you to Warramunga Seismic and Infrasound Facility, a critical observatory nestled in the heart of Australia's iconic Red Centre, with which I have an inseparable connection. This outpost stands as a sentinel, monitoring seismic activity and contributing to global efforts to ensure compliance with nuclear test bans. Through this chapter, I explore the importance of uniform spatial coverage by seismographs and why even the most remote and uninhabited places are vital for global seismology. As we navigate the harsh beauty of the Outback and its role in our understanding of seismic phenomena, we uncover the essential contributions of these isolated regions to the broader field of Earth science.

In chapter 9, we venture into the heart of a geophysical expedition to one of the world's most unwelcoming realms. The quest to deploy seismographs extends far beyond the red dust of the Outback, taking us to the floor of the Southern Ocean, to a spot positioned precariously between Tasmania and Antarctica. The chapter unfolds with accounts of a two-week quarantine in Hobart, followed by our voyage to the Macquarie Ridge Complex that separates the Pacific from the Australian tectonic plate. The ridge itself is an imposing underwater barrier that stands as the steepest of its kind on Earth's surface. Here, the fierce Antarctic Circumpolar Current, driven from the west and carrying the waves more than 10 meters high, collides with the ridge. This turbulent sea, where sailors have only the mythical sea nymphs to rely on, becomes the stage for our dramatic expedition. My research team faces some of the most extreme ocean currents, waves, and winds on the planet. Seasickness decimates the crew. Amid these harrowing conditions, the operation to place seismographs on the ocean floor becomes a test of endurance and resolve. The search for the MRO21 ocean bottom seismometer, which, after becoming uncontrollably adrift, adds an extra layer of challenge to the mission. The chapter paints a stark picture of the obstacles faced in the pursuit of science, set against the backdrop of one of the Earth's most unforgiving environments.

In chapter 10, we complete our journey, returning full circle to the origins of our story about planets and their formation. Driven by humanity's relentless desire to conquer new worlds, we set our sights on the Moon, guided by the heroic figures of astronauts Armstrong, Aldrin, and Collins from the Apollo mission. Their footsteps on lunar soil mark a pivotal moment in our exploration of space. From the Moon, we turn our gaze to Mars, the red planet that has long captured our imagination. Beneath its crimson dust, we seek evidence of ancient life and ponder the mysteries of its vanished magnetic field. Why did its magnetic field cease to exist? Is Mars's interior still geologically active?

We employ the most advanced techniques and commit our resources to uncovering these secrets. The InSight mission, equipped with a modern seismograph, embarks on a journey to Mars that far surpasses the capabilities of the Viking missions. Armed with the latest advancements in global seismology and our refined understanding of Earth's interior, my research group makes two significant contributions along with those of the InSight team and other researchers: First, the inferred Martian events that repeat in the same locations suggest that Mars's interior is still mobile. Second, by employing the waves that reverberate hours after they are generated by marsquakes and, with the help of mathematical tools, we confirm the existence of Mars's core. I should add here that I stopped writing in early 2022, and only a few edits were made for the material published after that date.

As we look to the future, we envision a time when the methods and knowledge we have perfected on our planet will be applied in missions to the far reaches of our Solar System. Our blue planet, the third marble from the Sun, will continue to be the foundation for our exploration—our home lab, as we carry the legacy of scientific discovery to new and distant worlds. I hope that *When Worlds Quake* will give you a glimpse of what that discovery might look like and, perhaps, inspire you to be a part of it.

1

Travelers through time

We live on planet Earth, the third marble from our parent star, the Sun, about 28,000 light-years away from the galactic center, in one of its secondary spiral arms.[1] Whoever named our galaxy the *Milky Way* probably spent many nights somewhere high in the mountains, in the darkness of a monastery or ancient observatory walls, staring at the night sky. From there, whiteness seemed to emerge from the sky's darkness, just like looking at a night path sprinkled with a bright liquid. The ancient Greeks attributed the phenomenon to the sleeping goddess Hera, who spilled milk while nursing Heracles. It is somewhat more romantic than the Godfather's Straw, as the Milky Way is called in some regions, after the godfather who stole a straw, and its contents fell out a little by little on the way and thus left a trail. Be that as it may, even before the invention of the telescope, ancient observers noticed that some objects that can be seen with the naked eye in the night sky behave differently from the planets and stars. They called these strange distant objects that they could not understand *nebulae*.

---

1. For definitions, see the glossary of relevant terms.

Later, when even finer details could be discerned with a telescope, it became clear that the nebulae are clusters of stars,
other galaxies far away from us, and the stars we see in the night
sky are just an integral part of our galaxy, the Milky Way. And
by no means is the Sun in a special place in it; we are even in a
minor instead of one of its four primary spiral arms. Over time,
the development of science and technology made the picture
less blurry and the details even more apparent. We then realized
that some of the stars, with their planetary systems, are forming
right now by the condensation of interstellar dust and gas into
massive mantles of ice, carbon monoxide, and ammonia.

The Sun is just one of the 100 to 400 billion stars in our
galaxy and just one of many other stars that, according to
today's knowledge of astronomers, have their planets.[2] Understanding the size of the universe and the fact that other stars
also hide planets in their bosoms while the number of discoveries about them continues to grow exponentially makes us think
about our own uniqueness. How could life on our own be different from life on other colorful specks studded on the black,
infinite canvas of the universe? Looking from this space perspective, the surface of our planet cooled down a long time ago,
but we can see from the countless volcanic cones, geysers, thermal springs, and shaking that its interior is still hot and in constant motion.

Indeed, earthquakes happen often as a result of the internal
dynamics of our planet. According to one catalog, in 2020
alone, worldwide seismographs recorded about 350,000 earthquakes, of which more than 200,000 were smaller than magnitude 2.0, which, due to their location or depth, escaped human

---

2. So far, more than 5,000 planets outside the Solar System (exoplanets) have
been detected, in more than 4,000 stars' orbits.

perception.[3] There are even many more, tiny ones, in places far from sensitive sensors that can record them—seismometers[4]— or deep in the Earth's lithosphere, making it impossible to detect them. Even if we consider only those earthquakes recorded by existing seismographs in 2020, we arrive at a somewhat frightening figure of at least a thousand earthquakes per day.

If the Earth were, in geological terms, a dead planet, if it did not have its internal dynamics, there would be neither plate tectonics nor a magnetic field, to name just a few of the global phenomena we first hear about during primary school education. We would be fried by radiation from space, from which the Earth's magnetic field separates us like a giant shield, and there would be no oceans or atmosphere favorable for life in the form they exist today. In short, life on Earth would not be possible. To explain the nature of the Earth in a little more detail, let's travel back to the very beginning of the formation of the Earth and the Solar System, in the geological eon Hadean, a little more than 4.5 billion years ago.

According to Greek mythology, Hades was the god of the dead and ruled the underworld. Twentieth-century geologists rightly felt that the period of Earth's creation, which began about 4.57 billion years ago, should be named after him because it resembled hell more than the birth of a new world. However, in that primordial chaos of the interstellar cloud of gas and dust, the building blocks of planets or planetesimals—objects larger than 10 kilometers—were gradually born, which gathered

3. We will say more about the earthquake magnitude in chapter 4.

4. Seismometer and seismograph are two terms that are often used interchangeably. However, in addition to the seismometer (sensor), the seismograph now also includes a digitizer—a device for converting ground motion data detected by the seismometer into digital data.

enough mass to begin gravitationally attracting other, smaller objects. With their mass and size, they gradually gathered the smaller parts remaining in the nebula, like a larger piece of plasticine with which a child picks up the crumbled remains on the surface of a table or linoleum floor after playing.

At the beginning of Hadean, the collisions of planetesimals, which would eventually form the Solar System as we know it today, were constant and violent. According to both the nebular hypothesis and mythology, from the primordial chaos arose Gaia, the poetic form of the name for Earth or the mother of all life. Gaia's eldest daughter, Theia, gave birth to Selene.[5] Indeed, no less spectacular than the creation of the Earth was the creation of the Moon and its cause-and-effect connection with the Earth, which is aptly described by the Greek names of planetesimals. Namely, when one of them, Theia, hit the proto-Earth, a good part of the volume of the proto-Earth was stretched, separated, and gradually solidified in orbit into a smaller object, the Moon. Its internal structure and rotation still bear the marks of that event.

Regardless of whether the Moon was formed by the collision of planetesimals or at the same time as the Earth, in that truly dramatic period, due to violent impacts, there was an additional increase in temperature and melting of materials in the interior of the proto-Earth. It most likely melted and differentiated several times so that the heavier chemical elements sank toward its center while the lighter ones remained in the mantle, crust, and proto-atmosphere. Several planetesimals have grown so much that they have picked up the remaining smaller, irregularly shaped bodies that until then were held together only by electromagnetic force. Although the growth dynamics of planetesimals are

5. Goddess of the Moon for the ancient Greeks.

still being understood with the help of computer models and simulations, it is clear that they were chemically stratified to the extent that there was iron in their centers and rocky silicates in their mantles. A surface layer (regolith) that did not solidify into compact rocks was bombarded by smaller bodies. This is the situation on the present-day surface of the Moon.

Remnants of primordial planetesimals, better known as comets and asteroids, are still around us. Fortunately, we can examine them directly, although we only recently managed to send and land research probes on some of them and bring samples home. For example, NASA[6] landed the probe *Near* on asteroid Eros in 2001, JAXA's[7] *Hayabusa* and *Hayabusa2* landed on asteroids Itokawa and Ryugu, and ESA's[8] *Philae* touched the surface of comet 67P in 2014. These missions attempt to contribute to understanding primordial material, organic molecules, and water by examining samples. In conference centers around the world, debates are also being held about the creation of the ocean, namely the possibility that water was brought to Earth later by comets.

But the most important thing for our story is that due to the kinetic energy (energy of motion) of the early collisions and the high concentrations of radioactive elements, a vast amount of heat was trapped in the Earth's center. It is still there today, about 3.8–4 billion years after the end of the period of bombardment of the Earth by smaller fragments that did not fit into it from the beginning. Due to high temperatures of around 6,000 degrees Celsius, the convection of molten iron is present in the outer, liquid core of the Earth. It, along with several other

6. National Aeronautics and Space Administration.
7. Japan Aerospace Exploration Agency.
8. European Space Agency.

ingredients, is crucial for creating and maintaining the Earth's magnetic field.[9] At the same time, convection also occurs in the mantle of the Earth, only on a much longer time scale, commonly termed the geological time scale. Although we have not yet been able to undeniably prove convection in the Earth's inner core, there are indications that the solid iron in the heart of our planet could also be susceptible to it due to the combination of viscosity, high temperatures, and pressures.[10] See figure 1.1 for the Earth's main shells.

The Earth's magnetic field is caused by the convection of electrically conductive iron in the outer core. As the Earth cools, it gradually crystallizes from the inside out; that is, the solid inner core grows from the liquid outer core. In addition to the heat released at the boundary between the solid and liquid phases of iron, the crystallization of the inner core causes a good part of the lighter chemical elements to rise through the liquid iron, like the crema in hot espresso that forms when air bubbles combine with oils from roasted coffee. In addition to the rise of lighter elements, additional turbulence occurs due to the Earth's rotation. This changing motion of the liquid metal ties to the electric current that generates the magnetic field. The electromagnet created in this way is often called the Earth's dynamo or geodynamo.

Although the magnetic field originates deep in the Earth's interior, magnetic field lines[11] pass through the interior, exit through the Earth's surface, and extend more than half a million kilometers

9. Convection in a liquid core transfers heat much more efficiently than thermal conduction, i.e., transfer of heat through mutual interaction of adjacent molecules. In general, we can say that a planet cools faster if there is convection in its interior.

10. See the publication by Jeanloz & Wenk (1988) in the bibliography.

11. Imaginary curved paths along which a magnetic monopole would move if we placed it in the Earth's magnetic field.

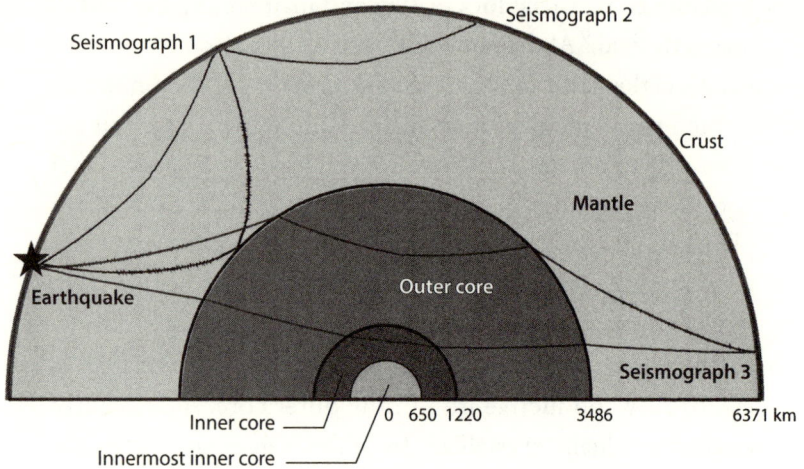

FIGURE 1.1. Cross-section of the Earth with a simplified internal structure divided into several main parts: crust, mantle, outer (liquid), inner (solid), and innermost inner (solid) core. Ray paths of seismic waves from the epicenter of the earthquake to the seismograph locations are marked for several types of seismic body waves: direct waves through the mantle of the Earth to seismograph 1 (so-called P or S waves), waves that are reflected from the boundary between the core and the mantle of the Earth and travel to seismograph 1 (so-called PcP or ScS waves), waves that are reflected from the Earth's surface at the location of seismograph 1 and travel to seismograph 2 (so-called PP or SS waves), waves that move through the outer and inner core of the Earth to seismograph 3 (the so-called PKP or SKS and PKIKP or SKIKS waves). Note: only P waves, unlike S waves, can travel through the outer liquid core. P waves in the outer core are denoted by the letter K.

into space, forming the magnetosphere. The solar wind and other radiation from space bounce off the magnetosphere or react with it in spectacular ways. One related optical phenomenon seen from the Earth's surface in the polar regions is the aurora. So, if there were no accretion of planetesimals in the Hadean, there would be no heat, and without it, no convection in the Earth's outer core and no magnetosphere that protects life on the Earth's surface.

In addition to the magnetic field, there is another, no less critical phenomenon directly responsible for life on Earth—plate tectonics. Tectonic plates, a dozen larger and many smaller fragments that make up the Earth's upper layer—the lithosphere—are in constant motion because the Earth is still warm in its interior. Still, they get stuck at the edges due to friction between their surfaces. Try to find two large blocks of packing sponge, press them together, and then move them laterally in opposite directions. At first, it won't work until you apply a force large enough to overcome the friction and move them. Even when the blocks finally start moving, the motion will be jerky and mostly slow at first, and eventually speed up, as if a slip had occurred. That's exactly how an earthquake is caused by sudden slippage on fault surfaces—boundary surfaces between entire plates or smaller fragments of the Earth's crust. However, due to the larger dimensions in nature, the larger the faults, the greater the possibility of a more significant earthquake. For example, the San Andreas Fault, which runs in the southeast-northwest direction and thus follows the west coast of America, has a plate the size of the entire Pacific on its western side and a plate much larger than the whole American continent on its eastern side.

In addition to convection in the Earth's outer core, we also mentioned much slower convection in its mantle. Imagine convection cells extending from the bottom of the mantle to the boundary with the Earth's crust, the way boiling soup circulates in a pot. Viewed in this way, the connection between the motion of the tectonic plates on the surface and convection in the mantle becomes much more intuitive. Namely, the tectonic plates float and move in the upper part of the convection cell as if on a conveyor belt. Of course, just because a planet is hot does not necessarily mean that plate tectonics is "operating," but that is a topic that goes somewhat beyond the scope of this chapter. Plate tectonics, which is manifest on Earth and which we will mention many times in the context of the main "culprit" for earthquakes on its surface, plays, however, a much more important role in the origin and existence of life on Earth, but perhaps in a bit more subtle way, than the magnetic field. Namely, plate tectonics regulates the amount of carbon in the atmosphere and oceans. To better understand why this is so, we must embark on another journey, at least in our minds. Instead of "flying" to asteroids and comets, we must immerse ourselves in the Australian Outback's magical red, where the oldest minerals in the Solar System have been found.[12]

If you thought diamonds were forever, you are probably not well-informed, since they can burn! However, if some material is indeed eternal, then it is the mineral zircon, an indestructible time capsule from the Hadean. Hadean scars on our planet's surface were long ago erased by the erosion and dynamics to which the Earth is subjected from within. Smaller traces remained only on rare, stable parts of the continent, for example, deep in the continental interiors of Siberia, Canada, and the Australian Outback. From the analysis of the zircons found there, we know that liquid water already existed 4.3 billion years ago, which means the Earth had a primitive atmosphere, but probably with so much carbon dioxide that life was impossible.

When the conditions for the beginning of plate tectonics were ripe, carbon dioxide began to enter the deep Earth by creeping under the tectonic plates in the places where they started to push each other. This process of advancing one tectonic unit

12. See chapter 8.

underneath another is called subduction. For example, the Pacific plate subducts under Japan and the Philippine plate at the east of the Philippines. On the other side of the ocean, it sinks under the western edge of the South American plate. The African plate is subducting under the Eurasian plate in the Mediterranean area, but, according to some new research, in the western part of the Mediterranean, the situation could change in the next million years so that the southwestern part of Europe starts subduction under Africa.

On the other hand, carbon comes to the surface through the formation of new oceanic crust in the mid-ocean ridge region, where hot material from the Earth's mantle rises and makes its way to the surface along the plate edges. In some places, hot material from the mantle makes its way to the surface and deep inside the plates in the form of volcanic eruptions. In those places—the so-called hot spots—magma finds its way to the surface in a spectacular way through a kind of underground pipeline and creates new islands, the most famous example of which is the Hawaiian archipelago, made of about eight larger and many smaller islands. By carbon recycling, deposition of carbonate sediments on the ocean floor, and re-emergence on mid-ocean ridges and in volcanic eruptions, the amount of carbon dioxide decreased enough to create conditions for life in the oceans and on the newly formed land. Without carbon recycling, Earth would be a greenhouse planet, just like its planetary sister Venus.[13] That is why the period in Earth's history when plate tectonics began is considered crucial for the beginning of life.

13. The greenhouse effect occurs when the gases in the atmosphere trap the Sun's heat. Venus has experienced the "runaway" greenhouse effect that prevented it from cooling and having liquid water on its surface.

Of course, the question remains whether plate tectonics needs to operate on a planet for life to develop. Perhaps life on Earth has evolved to adapt to the dynamic regime of our planet, and on other planets, it could adapt to their dynamic regimes.[14] Hydrothermal vents on the ocean floor are a source of ecosystem diversity. Erosion caused by plate tectonics results in the release of phosphorus, copper, and zinc into the ocean. They are the primary nutrients responsible for the existence of organisms such as plankton and, apparently, for the explosion of other living species, such as the one at the beginning of the Cambrian a little more than half a billion years ago.

In conclusion, the tectonic plates and their constant movement that causes strong earthquakes from time to time are tied by strong threads to life on Earth. How life is affected by the internal dynamics of the Earth, including the growth of the inner core and its effect on the magnetic field, and plate tectonics and its effect on balancing the proportions of the chemical elements, are some of the most profound questions of modern science. These two phenomena and events from Earth's turbulent past, the emergence of plate tectonics and the emergence of the geomagnetic field, are crucial for the development of life on Earth. Earthquakes are, therefore, kind of "underground giants" that exist as a consequence of the Earth's dynamics, "travelers through time" that have been there since the beginning. Dormant in the underground, they wake up periodically and send waves through the Earth's interior, disturbing the peace of every living creature on its surface.

---

14. We will return to this topic once more when we talk about why we place seismographs on the surfaces of other planets in chapter 10.

2

Namazu's tail
and Matilda's fate

In the darkness of the underground, deep in the mud beneath Japan, lay the colossal catfish *Namazu*. The god Kashima was holding him down with a huge oval rock and holding him still, but whenever something caught his attention or his grip loosened a little, Namazu would start to wiggle his tail and fins, and the Earth above him would start shaking violently. That is how, according to this myth, earthquakes occurred in the Land of the Rising Sun.

On 11 November 1855, when a great magnitude 7.0 earthquake struck Edo,[1] there were, according to historical records, about 10,000 dead, and more than 14,000 buildings were damaged. Up to 80 aftershocks continued to shake Edo daily, contributing to the psychosis and bitterness that accumulated in the people, who had already been distressed. Shortly after the earthquake, numerous woodcuts and prints (*namazu-e*) appeared depicting the Namazu catfish in all poses, most often

1. Today's Tokyo.

with the people whose stories had been circulating since ancient times that the catfish movement can be felt on the surface like an earthquake. The various paintings of Namazu reflect the emotions of fear, disgust and anger, so historians believe they reflect the social consciousness of the latest period of feudal shogunate rule.[2] Natural disasters are common catalysts that accelerate and bring to the surface the tensions that have been building up in society for years. The Edo earthquake was no exception, so it can be said that, in addition to the Earth's surface, the 1855 Edo earthquake also shook Japanese society's social and political foundations at that time.

Speaking of Japan, it sits right on the border between several major tectonic plates, and apart from picturesque volcanoes and cherry blossoms, its territory is crisscrossed by fault lines. From today's perspective, it is evident that it does not need Namazu or any other fish to cause earthquakes, which are just part of everyday life. However, the catfish figures in this story for another reason: according to some myths, fish are among the animals that can predict earthquakes. However, decades and even centuries have passed, and much water has flown under the Shirakawa[3] bridges, and scientists there cannot confirm the connection between catfish behavior and earthquakes. Instead, Japan relies on sophisticated early warning systems that recognize the onset of an earthquake and send messages to humans and computer systems, which control traffic and momentarily slow and stop high-speed trains before the devastating shear waves arrive. All of this is possible and makes sense in a country with such a high

2. The three main periods during the shogunate, from 1192 to 1868, were Kamakura, Ashikaga, and Tokugawa. In 1868, the emperor regained power, and the period that followed is referred to as the Meiji Restoration.

3. A river that runs through the Gion district of Kyoto.

population density,[4] especially when you know that even a few seconds are enough to avoid a disaster.[5] Figure 2.1 illustrates excellent coverage of seismographs in Japan.

In addition, the Japanese rely on prevention, which means that they strictly adhere to construction according to building regulations. I experienced earthquakes several times during my stay in Japan, in my tiny apartment on the fourth floor of a building in Yokohama and at the JAMSTEC[6] building at my workplace. During an earthquake, one feels like one vibrates on springs. To put things into perspective, there will be almost no damage in Japan during an earthquake like the magnitude 5.3 one in January 2020 that hit the Croatian capital, Zagreb, or magnitude 5.4 several months later, damaging historical buildings in Ponce of Puerto Rico. Building codes in Japan have changed five times since 1971, with perhaps the most drastic change after the 1995 Kobe Earthquake. Indeed, walls are installed inside buildings that dampen seismic movements, and absorbent materials under the foundations are also used., for example, laminated rubber, springs, and ball bearings that block ground motion quite well when seismic waves generated by earthquakes pass through.

One of the first modern devices for measuring ground motion (a seismograph) was constructed by John Milne, an English mining engineer employed by Imperial Japan as a foreign advisor and professor of geology and mining in Tokyo. Interestingly, because of seasickness, he traveled by land for several months via Siberia and then stayed in Japan for 20 years. The Japanese Seismological

4. It amounts to 350 people per square kilometer.

5. Japan's seismographic networks are impressive. Figure 2.1 shows one of these networks with high density across the Japanese archipelago.

6. Japan Agency for Marine-Earth Science and Technology.

FIGURE 2.1. Top: A display showing several different seismic networks operating in Japan by the National Research Institute for Earth Science and Disaster Resilience (NEID) at its headquarters in Tsukuba City, Japan. The High-Sensitivity Seismograph network, Hi-net, is the most impressive permanent seismic network, comprising nearly 700 seismographs (https://www.hinet.bosai.go.jp/). Apart from that, NEID operates a nationwide broadband seismograph network known as the F-net. Satoru Tanaka (who took the photo) and I visited the headquarters during my stay in Japan, supported by the Japan Society for the Promotion of Science (JSPS). Subsequently, we used these networks' waveform data to research the Earth's deep interior. Bottom: A focus on Honshu, the largest of the Japanese islands, and the location of the seismographs.

Society financed the construction of the first seismograph, on which Milne worked with two other collaborators, also English. Today, Milne is credited with the construction of the first modern seismograph, completed in 1880, and according to some, he is considered the father of modern seismology.

His seismograph consisted of a mass suspended on a horizontal arm, attached to a vertical rotating axis, and suspended by a silk thread. At the end of the arm was a horizontal plate with a narrow opening parallel to the arm; below it was another parallel plate with a narrow opening perpendicular to the upper opening. That bottom plate was placed on top of a box containing a drum of rolled-up, light-sensitive photographic paper. For the mass to move during an earthquake, the oil lamp light passing through the slits in the tiles would emulate the writing of pencil on paper. Thus, at the end of the 19th century, the first records, seismograms, were created in a dark room at the University of Tokyo. Of course, today's seismographs are of the electromagnetic type, and recording is achieved by the movement of a conductor through a magnetic field, during which the electric current is recorded, and the data are digitized. Figure 2.2 shows an original analog seismograph preserved.[7]

———

On the fatal day of 18 April, far back in 1906, Matilda the cow patiently waited for the early morning milking on a farm on the Point Reyes Peninsula, about an hour's drive north of San

7. I took this photo during my visit to the GeoForschungsZentrum (GFZ) in Potsdam. More about Wiechert seismographs and even a 24-hour "live seismogram" can be found on the website of Mohorovičić's memorial rooms, at http://www.gfz.hr/sobe-en/seismographs.htm.

FIGURE 2.2. Top: A display of a Wiechert seismograph at the Research Center for Geosciences (GeoForschungsZentrum; GFZ) in Potsdam, Germany. GFZ is surrounded by world-famous scientific institutes for astrophysics, meteorology, geodesy, and geomagnetism, funded in the second half of the 19th century. The seismic station started operating in 1902, and the 1906 San Francisco earthquake was recorded with a Wiechert seismograph.

Francisco. Dense fog at dawn is common here because the cold oceanic air mass begins to move toward the warmer interior of the North American continent in April. Point Reyes was given its name by Spanish maritime explorers of the 17th century because, from the deck of a sailing ship, its magnificent appearance strongly reminded them of the place where the kings ruled—*la punta de los reyes*.[8] The protruding part of the land, covered with green pastures, rises almost half a kilometer above the ocean surface. The narrow bay of Tomales separates it from the rest of the California coast like a giant zipper. Lashed by the waves of the Pacific Ocean, from the air, it looks like a large piece of land that has partially broken off and sailed toward Alaska.

With its appearance, Point Reyes must have evoked the sigh of at least one sea wolf with rough facial features, eroded by salt and wind, who in those years ventured on a long voyage across the Atlantic and sailed around Cape Horn along the western coasts of the two Americas all the way here. Shrouded in fog, that piece of land looks mysterious, just enough to become an indispensable location for Alfred Hitchcock in *The Birds*[9] and John Carpenter in *The Fog*. After all, I couldn't resist the charms of Point Reyes, even myself, in the days when I worked as a graduate student at Berkeley[10] and later as a postdoctoral fellow at the Lawrence Livermore National Laboratory.[11] My faithful 1973

8. From Spanish, the place of the kings.

9. *The Birds* was filmed just a little north of the Point Reyes peninsula in the small town of Bodega Bay.

10. By its full name, the University of California at Berkeley was founded in 1868 and is the oldest university in the state of California university system, of which there are ten today.

11. Lawrence Livermore National Laboratory is one of 17 national laboratories in the United States of America. It is known for leading research work and development of science and technology for application in the field of state security, and

Ford Maverick must have blended perfectly with its green color every time I drove it from the San Francisco Bay Area to this special place for me.

When, after a long winding road, you finally reach the most prominent point of the peninsula, you are instantly rewarded by the view from the top, from the shade of centuries-old cypress trees, whose crowns have been shaped into arcades by the winds over the years. The view to the north is to one of the longest and most impressive sandy beaches in Northern California.[12] On the western side, 308 steps descend almost perpendicularly to the famous lighthouse, whose silhouette merges with the ocean. With some luck, you will see a grey whale on the high seas, grabbing toward the north.

The cow Matilda could not have imagined that deep in the rocks under her feet, a drama was unfolding that would permanently change our understanding of earthquakes and shape several scientific disciplines that were just being born at the beginning of the 20th century and had progressed mainly in Europe and Japan until then. Beneath Matilda, the tectonic plates, which had been slowly moving, pushing, and straining the rock mass of Point Reyes for decades, were now so strained that it was not a question of if, but only when, they would rupture and send seismic waves in all directions. But who could have guessed the existence of tectonic plates then? Who could have known that on that very day, San Francisco would experience an earthquake of magnitude 7.7,[13] and three quarters of the city would tragically end in fire?

---

seismology as a scientific discipline is the main competency for discrimination of nuclear explosions from earthquakes.

12. The Point Reyes beach is 18 kilometers long.

13. See chapter 4 for more on earthquake magnitudes.

It was early morning, about 20 minutes before dawn. At that moment, just after 13 minutes past 5, the rocks squealed and creaked loudly, and the underground architecture gave way and failed somewhere. The eerie sound echoed through the misty hills along the ocean, the forests of thousand-year-old redwoods, and the densely arranged vineyards of Sonoma. The first impact arrived about 20 seconds before the main effects. The main shock violently shook the entire coast for more than 40 seconds, and when it was over, it became clear that Point Reyes had moved more than 8 meters along that great zipper to the north-northwest, just like in the title of one of Hitchcock's movies.

One of the eyewitnesses at Point Reyes later told reporters that the Earth opened just below the unfortunate Matilda, separated her from the other cows, and swallowed and buried her so that only her tail was left sticking out of the crack. When I visited that farm for the first time with other graduate students, Mark Richards showed us the hayloft and barn that slid in one direction on the Pacific Plate. A large pile of manure slid in the other direction, leaving only a dark stain on the outside wall of the barn. The large zipper the plates were sliding along is actually— you guessed it—the San Andreas Fault. This fault, as we know very well today, follows the California coast with a length of about 1,300 kilometers. The Point Reyes Peninsula is part of the Pacific plate on the west side, and the American continent is part of the North American plate on the east side of the fault.

Be that as it may, this is how the myth of Matilda was born. It turns out that by sheer coincidence that morning, she had two feet firmly on the Pacific plate and the other two on the North American plate. Many tourists today who come to see the famous earthquake path near this farm try to emulate Matilda by posing for photos. So, they jump excitedly from one plate to another, momentarily forgetting the unfortunate fate of the famous cow, the agony of San Francisco, and the turbulent history of this place.

Stories and myths about the Earth opening and swallowing animals, people, houses, or entire cities have been told for centuries. In those days back in 1906, they appeared with greater frequency and in numerous newspapers worldwide. Of course, from today's perspective, if the Earth indeed just opened and closed, there would be no friction or stress and, ultimately, no earthquakes as we know them. People believed that the faults that could be observed on the surface of the Earth were the result of earthquakes; that is, they did not exist before the earthquake happened. The continuation of this story will reveal how delusional they were.

———

Just a few hours after the earthquake, Jack London wrote in his famous article "The Story of an Eyewitness": "San Francisco is gone. Nothing remains of it but memories and a fringe of dwelling-houses on its outskirts."[14] Due to the rupture of gas pipes, a fire broke out somewhere around Montgomery Street in the business district and south of Market Street and spread to Russian Hill, Chinatown, North Beach, and Telegraph Hill. More than 500 blocks of charming and sometimes quaint Victorian houses on a dozen or more square kilometers were engulfed in flames that burned and glowed for four days and nights. About 3,000 people lost their lives, 28,000 houses and buildings were destroyed, and about 300,000 people had to seek refuge in refugee camps set up in Golden Gate Park and the surrounding areas of the city, which was three quarters of

14. See London (1906) in the bibliography.

the then population of San Francisco. Indeed, when four days after *The Earthquake and Fire*, the last sequoia beam was burned, and the spring rain soaked the ashes, the terrible consequences became evident, and Twain's words echoed painfully and heavily. The city was facing new challenges. However, it was built even more beautifully and more modern before the Panama-Pacific International Exposition in 1915.

Who else but Berkeley geology professor Andrew Lawson was chosen to lead the earthquake research commission? The choice was logical: Lawson was the first professor in America to create a geological field course and probably discovered the San Andreas Fault on one of those fields around 1895. Mrs. Lawson was once asked what her husband's religion was, and she replied simply, "He's a geologist." I don't know what he might have thought about the earthquakes that hit San Francisco during the 19th century, and there were some, although not that big, and whether he connected them in any way with the existing fault.

Launched shortly after the earthquake in April 1906, it was the first organized commission of inquiry into earthquakes in the United States. It comprised 25 geologists and astronomers from several leading American observatories and universities. Its members walked and rode kilometers and kilometers along the fault zone and reported in detail about the movements around the fault and the resulting damage. They measured everything that could be measured: how and to what extent the walls, roads and paths, canals, fences, rows of trees, orchards, vines, and roots of old trees had moved. Although the coastal sequoia,[15] with its impressive height and red-colored trees, was a widespread native species, by 1906, the Franciscans had

---

15. *Sequoia sempervirens* or redwood, as opposed to *Sequoiadendron giganteum* or giant sequoia that grows in the Sierra Nevada.

already brought European apple trees and cypresses from Europe and eucalyptus from Australia. Among the vineyards of Sonoma and Napa, which were planted even before the gold rush, there were, in addition to Spanish, French, and Italian grape varieties, a lot of Zinfandel.[16] Undoubtedly, some shifted Zinfandel rows ended up as data in the notebook of one of Lawson's colleagues.

According to Lawson's report and subsequent analysis, about 470 kilometers of the northern part of the San Andreas Fault ruptured. As a rule, an earthquake starts from one point—the focal point or hypocenter. The rupture then spreads from it—depending on the state of tension in the Earth's lithosphere—either in one direction along the fault or bilaterally. In this case, it started from a focal point under the ocean, about two miles west of San Francisco, and spread northward, past Point Reyes and Fort Ross to Cape Mendocino, and southward to San Juan Bautista. In fact, it is unknown how long the fault ruptured in the north because somewhere near Point Arena, it turns westward and disappears under the Pacific Ocean. By the way, the speed with which the rupture spread was a terrifying 3 to 3.7 kilometers per second, ten times faster than sound in the air, and so 15 times faster than a Boeing 747.

Lawson's selected team of experts found an excellent correlation between the resulting damage and the type of soil—namely, the buildings on unconsolidated sediments in the city and its surroundings suffered the most significant damage. The least

16. It turned out that the Zinfandel variety is genetically identical to the indigenous grape variety Tribidrag from the Adriatic coast. Crossing Tribidrag (also known as Crljenak Kaštelanski) with the Dobričić variety created the most important contemporary red wine variety in Croatia, Plavac Mali. Zinfandel was a real hit among immigrants from the Mediterranean and Russia and later became a hallmark of Californian viticulture and winemaking.

damaged area was on the surrounding hills near the houses built on solid rock. In the report, the displacements along the fault were described in detail. It turned out that the largest were near the fault and that their size gradually decreased in the direction transverse to the fault. The largest ground movement was 8 to 9 meters at Point Reyes. In addition to the main one, they found many parallel faults called the San Andreas Fault system. Each has the potential for devastating earthquakes, perhaps most notably the Hayward fault, which runs parallel to the San Andreas on its eastern side and through Berkeley and Oakland. Berkeley's largest sports stadium, which had to be built on the Hayward fault in 1922, is constructed of two halves connected by expansion joints. The relative motion of the two halves is about half a centimeter per year.

The observations described in the report served as the basis for the elastic rebound theory developed by one of the participants in the campaign, Harry F. Reid of Johns Hopkins University. It followed the previous work of Grove K. Gilbert from USGS,[17] also one of the participants in the fieldwork. Reid had made observations from the field before and after the earthquake. For example, he observed a fence that had slowly deformed before and then broken at the moment of the earthquake. In the beginning, the hypothesis, and then the theory, described what every geophysics student knows well today: that tension gradually accumulates in rocks and deforms them until it reaches the actual strength of the rock. When this happens, the rocks suddenly snap or spring like a hair band when you pull them in two directions and return the tension to the initial settings. This rebound is known to us as an earthquake;

17. United States Geological Survey.

in it, the energy is suddenly released and travels in the form of seismic waves through the Earth and its surface.

Faults and earthquakes go hand in hand with one another. They have been surely present in California for about 30 million years since the Pacific and American tectonic plates met for the first time and formed the San Andreas Fault zone. However, it was not recognized until 1906 that earthquakes occur on existing faults; the opposite was thought—that earthquakes lead to faults. Although plate tectonics as a theory languished in the bottom drawer of geophysics for another half century, Reid recognized that the forces that deform the rock masses on one and the other side of the fault must be somewhere far away and not in the vicinity of the earthquake's focus.

3

Moho's and Inge's binoculars

Because he was born in 1857 in Volosko, a small town on the coast of the Adriatic Sea (fig. 3.1), it can be assumed that the young Andrija Mohorovičić grew up surrounded and fascinated by the sea. In its calm moments, the sea generously filled the fishing nets, but it frustrated fishermen with its cruelty in the hours when the stormy *bura* wind alternately pushed and swatted masses of cold air from the direction of the Pannonian Plain, often reaching speeds of 200 kilometers per hour. From his family house built on white limestone blocks, which even today stands only a few meters from the sea, Andrija watched seagulls, clouds, and stormy seas at the age when he was deciding on his life's calling.

When Andrija was only one year old, Darwin's book *The Origin of Species* was published, and while Andrija was sitting in school as a first grader, Jules Verne's novel *Journey to the Center of the Earth* was published. Who knows when Verne's work became mandatory reading, but in it, the Earth's center is described as a vast hollow filled with a subterranean ocean on whose shores giant mushrooms grow, and in its murky waters, ichthyosaurs

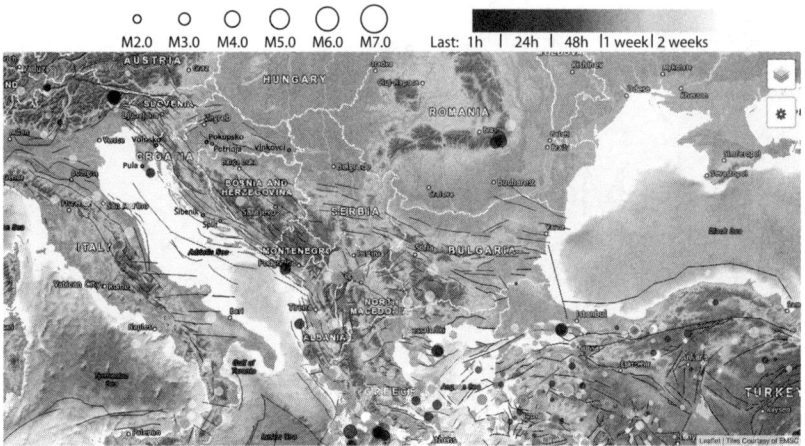

FIGURE 3.1. Two weeks of seismicity in the Mediterranean in January 2025, with most earthquakes on the Apennine and Balkan peninsula (notably, in Greece) and Anatolia (Turkey). Only magnitudes 2.0 and above are shown, with the size of the circles proportional to the magnitude. Major faults are displayed with black lines. The most prominent fault is the North Anatolian Fault in Turkey, running roughly east-west. Note many smaller faults and their orientation crisscrossing the area. The localities of the cities and towns in Croatia and the surrounding countries mentioned in the text are also shown on the map. The colors show the chronology of the earthquakes. This map is produced using the catalog and mapping tools from the Euro-Mediterranean Seismological Centre at https://www.emsc-csem.org/.

and plesiosaurs clash in the struggle for survival. At that time, very little was known about the Earth's interior, and people inclined to vertical travels, like Professor Lidenbrock in Verne's novel, whose only wish was to descend to the center of the planet on a magical slide, were rare.

Have you ever thought of geophysicists as people inclined to such vertical thoughts, and have you wondered what drives them? What goes through their heads? What unusual or mysterious phenomenon or unsolved puzzle can control their emotions and put a smile on their faces even before the morning coffee? Something so glorious that it can ground them in an instant or turn them into weirdos who, in their rudeness, can walk past you without even noticing you? Is it the part that refers to *geo, physics,* or both? Do they look forward to earthquakes and volcanic eruptions, or do they dread them just like the rest of the ordinary world? What exactly has the power to motivate a geophysical mind in today's visible and tangible world, knowing that many problems in geophysics deal with phenomena or objects that are kilometers below our feet? Knowing that the answers are crouching deep in the darkness, beneath the Earth's shells, like a naked body covered by layers of clothing hiding from the prying eyes? And finally, knowing that we don't have a magical slide to descend to the depths of the Earth?

That is why our story begins with Andrija Mohorovičić, a geophysicist who once said, "The task of seismology is to study the interior of the Earth and to continue where geology stops, and seismographs are like binoculars—devices that allow us to look into the deepest layers of the Earth."[1] What did he want to tell us with this sentence, and how did he manage to do precisely what he did? What did he discover after all? To answer these questions, let's take a short trip to his world at the end of the 19th and beginning of the 20th centuries, at least in our minds.

After finishing high school in the city of Rijeka, Andrija said goodbye to his family and, like many of his contemporaries, started a new chapter in his life as a student at one of the most

---

1. This story has already been told to an audience of geophysicists and geologists, who are most naturally interested in hearing it. It was printed in a slightly abridged form in Japanese in *Naifuru*—the journal of the Japanese Seismological Society—after great interest from colleagues in Japan on the occasion of the centenary of the discovery of the Mohorovičić Discontinuity in 1910. See Tkalčić (2016) for more details.

prestigious Austro-Hungarian universities—Charles University in Prague. There, he studied physics and mathematics at the Faculty of Philosophy.[2] Professor Mach, after whom the unit for the speed of sound is now called, taught him physics. During his studies, Andrija also took the theory of elasticity, which describes how a material deforms when a force is applied. The theory of elasticity later proved the key to understanding the propagation of waves from the earthquake's focus through the Earth to the instrument (seismograph) on its surface that records this motion.[3]

Thus, our 20-year-old Andrija, even before the first modern seismograph was constructed and the waves of distant earthquakes were recorded on it, unwittingly put himself in an excellent position that one day he would make a great discovery in geophysics. But he achieved that 30 years later. In the meantime, he established himself as a meteorologist, and seismology was born as a new scientific discipline.

When Andrija returned to his homeland from his studies, he first worked in secondary schools in Zagreb (1879–1880) and Osijek (1880–1882) and then taught mathematics, physics, and meteorology at the Nautical School in Bakar (1882–1891). Those were exciting years of the 19th century, each of which spawned at least a few patents without which life today would be unimaginable. For example, the light bulb was patented in 1879, the cash register for payment in 1884, and contact lenses and the gramophone in 1887.

---

2. In Klementinum, where the University was once located, the Czechs erected a memorial plaque in honour of Andrija Mohorovičić (fig. 3.2).

3. In the first approximation, the solid Earth behaves as an elastic medium during the passage of seismic waves, i.e., the deformation is proportional to the force. This law is also known as Hooke's Law, after the English scientist Robert Hooke.

FIGURE 3.2. The bust of a commemorative plaque to Andrija Mohorovičić, in the Klementinum in Prague, Czech Republic, where he was a student. The bust is designed symbolically, showing the two layers of the skull.

It is obvious that young Mohorovičić was not only occupied with clouds, because, at that time, he married Silvija Vernić, with whom he had four sons: Andrija, Ivan, Stjepan, and Franjo. That was when he began his scientific work in meteorology, which culminated in 1893 with his doctoral dissertation: *The Results of the Observation of Clouds in Bakar.*[4] As one of those most responsible for establishing the Meteorological Institute, today known as the Geophysical Institute, he worked there until 1922. Explanation of the so-called atmospheric rotor is his

4. Three of his previously published papers were recognized as an equivalent of a doctoral dissertation. See the publication on Mohorovičić by Orlić (2019) in the bibliography.

most famous work in meteorology. Still, it is perhaps less known that he took upon himself the responsibility of publishing the weather forecast and did it every day for the next 20 years! Therefore, it can rightly be said that Mohorovičić is the father of weather forecasting in Croatia. It is not entirely clear what exactly led the then relatively successful middle-aged meteorologist to shift his attention from meteorology to seismology. It is possible that he was interested in earthquakes in the Mediterranean. Figure 3.3 shows 10 years of earthquakes (between 2012 and 2022) with magnitudes larger than 2.5 in a part of the Mediterranean Sea. The map is dominated by the Apennine and Balkan peninsula seismicity. After John Milne and some European physicists constructed the first modern seismographs in the last quarter of the 19th century, it was precisely the earthquakes from the Mediterranean that were most clearly registered on seismographs in Europe, because the rest of the earthquake areas, for example, Japan, Tonga, Fiji, New Zealand, Alaska or the west coast of America, were much farther away. From there, only great earthquakes could produce strong enough waves to be recorded in Europe with the technology of the time.

When the middle-aged Mohorovičić borrowed a seismograph from Budapest and set it up for work in April 1906,[5] he had no idea that one of the first earthquakes he would record would be the famous one from San Francisco. As we have seen,

5. Interestingly, in 1906, the Irish seismologist and geologist Richard D. Oldham interpreted the existence of the Earth's core from the seismological analysis of P waves. German-American seismologist Beno Gutenberg confirmed Oldham's observations in 1913 by determining the discontinuity depth to be 2,900 kilometers. It was not until 1926 that academic circles accepted that the nucleus was in a liquid state. See publications by Oldham (1906), Gutenberg (1913), and Jeffreys (1926) in the bibliography.

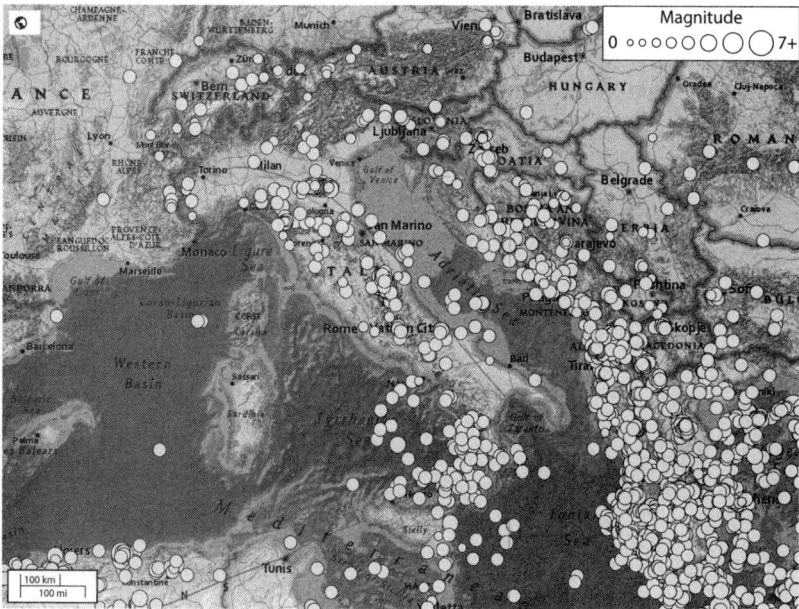

FIGURE 3.3. Ten years of seismicity (2012–2022) centered on the Apennine and Balkan peninsula. Only magnitudes 2.5 and above are shown, with the size of the circles proportional to the magnitude (see the legend). Note that most larger earthquakes are overlaid by the subsequent smaller aftershocks, so they cannot be deciphered. Major tectonic boundaries are displayed with red lines. This map is produced using the catalog and mapping tools from the United States Geological Survey (USGS) at https://earthquake.usgs.gov/earthquakes/map/.

this earthquake on the San Andreas Fault, which was followed by a fire and the destruction of San Francisco, brought a new paradigm and sparked a new scientific discipline, seismology. Namely, the commission that investigated the earthquake realized that earthquakes do not cause the formation of faults, but on the contrary, new earthquakes occur on faults that already exist. This may seem relatively trivial from today's perspective but don't forget that at that time, there were still stories about the earth opening and swallowing a grazing cow, someone's property, or even an entire street, and a new fault would form in that place. Remember, before the San Francisco disaster, there was a deep-seated belief that earthquakes could happen anywhere and anytime, but the new paradigm truly changed how we thought about them. Perhaps this new knowledge influenced Mohorovičić and his decision to devote himself entirely to seismology.

During a sudden slip or earthquake, the energy has accumulated for months, years, and sometimes even centuries and is suddenly released as waves. When these waves pass through the Earth, the oscillation of the ground particles can be either in the direction of the waves, just as air molecules oscillate during sound, or perpendicular to the direction of the waves, more like the waves on the surface of the sea. The first type is better known as P waves (Latin *primae*), and the second is S waves (Latin *secundae*). P waves always precede S waves when they reach the Earth's surface from its interior. The sensitive pen of the seismometer recorded their motion on a moving piece of long paper wrapped around a rotating mechanical drum (see a similar mechanism in figure 2.2).[6]

Over time, Mohorovičić learned to recognize P and S waves from other sounds recorded on the seismogram and to distinguish Mediterranean earthquakes from those from the Far East, just like an experienced fisherman who knows in advance which fish it is by the twitch of the hook. From the time that passed between the arrival of the P and S waves, he could deduce how far away the earthquake was, using the same logic he used as a boy to determine the distance of a storm cloud by counting the seconds between the flash of lightning and the sound of thunder.

6. The first records were analog, and digital ones later replaced them.

From the intensity of the recorded signal, he knew how big the earthquake was.

Every morning after arriving in his office, he impatiently went down to the room where a lone seismograph listened to the Earth's sound at night, and Mohorovičić thought about what he would see there. Sometimes, it was nothing special, and sometimes, he was greeted by a true symphony of signals. Every vibration of the ground and every recorded seismogram was part of a puzzle that, like a fingerprint, revealed the mechanism of the earthquake and the image of the Earth's interior. The more seismograms you had, from as many corners of the world and from as many different distances from the earthquake as possible, the easier the task was. In terms of how thorough Mohorovičić was, to the point that even the most minor details failed to escape his attention, he was unrivaled. That's precisely why he was struck by what he observed on the seismograms of the Mediterranean earthquakes: there were some other signals present in them that, according to the knowledge of the time about the propagation of waves through the Earth, no one expected.

———

In the morning hours of 8 October 1909, the sudden mooing of cows echoed through the Kupa River valley. The waves from the earthquake had set off on their long journey through the Earth's interior and first violently shook and devastated the rural houses and barns there and then continued their movement through the surface and interior of the Earth, reflecting and refracting from the invisible underground architecture just as light is reflected and refracted at the water surface.

When the earthquake struck that morning from the direction of Pokupsko (see figure 3.1), about 35 kilometers south of

Zagreb, Mohorovičić knew that, to piece together the puzzle of that earthquake successfully, he had to immediately request seismograms from his colleagues from other European cities who had seismographs at the time. When, after several months, he finally opened the last envelope that arrived at his address, he was amazed by what he saw on the seismograms, and an idea dawned on him.

In principle, there are two ways of approaching a problem: deductive, when you start from a general law that you know well and then try to explain observations based on that, and inductive, when you begin from observations and try to find a general law or phenomenon that leads to these observations. The first is known in geophysics as the "direct method," and the second as the "inverse method," which is something like a piece of everyday bread for every geophysicist. In fact, the inverse method is also known as the inverse theory. It is crucial in the exploration industry because, from our observations on the Earth's surface, we need to infer something about its interior, for example, whether there is oil somewhere and whether we should drill in that place or not. To generalize this to global geophysics and our story about Mohorovičić, the inverse method is critical because it contributes to the efforts to infer from the seismological observations the structure of the Earth's interior, from the surface to its center. For example, in the late 1960s and early 1970s, seismologists "borrowed" the idea of computer tomography from medicine and developed it into seismic tomography with the aim of "recording the Earth's internal organs" as best as possible.[7]

It is important to think briefly about this inverse theory and how it works before we delve into deeper thinking about how

7. We will return to this in chapter 6.

Mohorovičić came to his great discovery. Namely, in science, success is often measured by how accurately we can explain or predict an observation from nature, space, or the laboratory by knowing natural laws. The set of parameters we use to describe a problem is called a "model." If our understanding of natural laws is correct and if we have chosen the parameters well so that our model is good, then we will succeed in the goal of predicting the results of observations well, and we can righteously say that "the fit of the observed data to the predictions is good." If the model misbehaves somewhere, the observation forecast is not good either, indicating at least two possibilities: either we misunderstood the natural law, or we chose the wrong parameters to describe it.

If the above sounds too abstract, let's take, for example, the seismogram from Vienna that Mohorovičić received after the Pokupsko earthquake. We can know the exact location and depth of the earthquake, the exact time when it occurred, everything about how the energy was released and spread in the form of waves through the homogeneous Earth from the focal point to the seismometer, and, no less importantly, understand how the seismometer's pencil records the shaking of the ground on a piece of paper—but all this is still not enough. That is because the Earth is not homogeneous in its interior. That is why we need to know the fine details of its structure between the focus of the earthquake and the seismometer. Hence, what materials it is made of, whether there is an underground boulder in the way of the waves through the Earth's interior, how big it is and what its shape is, what its density is, and what is the speed of propagation of waves through it, etc. If we know all this perfectly well, we can insert these data into computer code and synthesize a digital seismogram. When we compare this synthetic seismogram of ours with the one recorded in Vienna, we will either jump to the ceiling with joy or go back to the drawing board.

Mohorovičić's reaction was neither of those two. In fact, he had no way to calculate a synthetic seismogram at the time. He had a pretty good idea of the location of the earthquake based on many observations, he could make a good guess as to how the energy was emitted, and he had excellent knowledge of seismometry.[8] But he didn't know the fine details of the Earth's structure and exactly how it modulates waves as they pass through it. At that time, it was assumed that in addition to the core, discovered in 1906, the Earth also has a mantle on which the continental and oceanic crust could float, but this was only a guess.[9]

Let's imagine that a magnitude 5.5 earthquake occurred in an unknown location in Europe, and you have three seismometers: one in Madrid on the Iberian Peninsula, another in Athens, Greece, and the third in Paris, France. Let's also imagine that you were lucky enough to recognize the onsets when the P and S waves arrived on the recorded seismograms, with the note that in practice, this is not always so clear because some sensors, due to their location, may be more susceptible to noise than others. The noise is created by the sea, the ocean, and the atmosphere, which interact with the solid earth and the anthropogenic activity caused by people. In Athens, you measured on the seismogram that the time elapsed between the occurrence

8. The theory of how a mechanical seismometer records ground oscillations.

9. Back in 1897, the German geophysicist Emil Wiechert proposed a model of the Earth according to which it has an iron core and a silicate mantle. In 1912, the German meteorologist Alfred Wegener proposed the continental drift hypothesis, according to which the continents change their positions on the Earth's surface. It would later be rejected due to the lack of evidence that the continents can "plough" through the ocean floor under the influence of tidal centrifugal forces and was replaced first by the hypothesis and then by the theory of plate tectonics.

of P and S waves was 85 seconds; in Paris, the difference was 125 seconds, and in Madrid, 147 seconds. Remember the example of lightning and thunder here, except that the speed of sound through air is only 0.34 kilometers per second, and the speed of light is practically infinite compared with the speed of sound, so 3 seconds between lightning and thunder would correspond to a distance of 1 kilometer. Even without calculations, if you remember that P waves move faster than S waves and use an analogy with the example of lightning and thunder, it is immediately apparent that the earthquake is closest to Athens and farthest away from Madrid. Assuming that P waves move at an average speed of 8 kilometers per second and S waves at an average speed of 4.5 kilometers per second, we can calculate, based on the time differences between the occurrence of S and P waves, that the earthquake is 875 kilometers from Athens, 1,285 kilometers from Paris, and 1,512 kilometers from Madrid.

We determined that the earthquake was 875 kilometers away from Athens, but it could have occurred in any direction from Athens. In other words, the solution is somewhere on a circle with a radius of 875 kilometers centered in Athens, maybe somewhere in the sea, and maybe on land. Suppose we apply the same logic to Paris and Madrid and draw the other two circles with centers in Paris and Madrid on the map. The circles will intersect approximately in one point that belongs to all three of them.

It turns out that this point, our solution, is the city of Naples. If I described this well, you understood the principle of the triangulation method! Apart from determining earthquake location, it is also used in astronomy, geodesy, and military purposes. However, it is more important for our story that the triangulation method is an example of an inverse method, where with certain assumptions and from the measurement of the time difference between the arrival of S and P waves on seismograms, you are able to determine the location of the source of the earthquake.

From the above example, we learned the principle of the triangulation method, but are our assumptions always correct? First, we assumed that everything takes place in a plane and not in space; that is, the earthquake's source is a point on the Earth's surface, which is rarely the case. Second, we assumed that P and S waves propagate from the earthquake source to the seismometer across the Earth's surface and that their average speed is known and equal in all directions, which is also not the case, because P and S waves move through the Earth's interior, which is inhomogeneous. That changes their speed, the same way your body changes speed when running at the same power as you move from sand to a solid surface. Translated into the language of geophysicists, there are errors in the theory and the model, which means that a direct calculation of wave travel times will be wrong. It follows that the inverse method will not result in a good fit for the observed data. In other words, the earthquake location calculated this way will be inaccurate.

Today, we know what the paths of P and S waves look like through the Earth's interior. From tomographic studies of the underground, we can calculate the speed of the waves in various places, which means that modern direct and inverse methods used to determine the location of earthquakes are far more accurate than in previous years. We have more seismometers, meaning that we can use many more of them than three, as in our example. However, there is still an error of about a few kilometers, and in places where the structure of the Earth is less known—of more than ten kilometers.

After this excursion into theory, let's get back to Mohorovičić and what occupied him then. If he tried to calculate the location of the Pokupsko earthquake in the previously described way, and perhaps even corrected for the depth of the quake and assumed that the paths of the waves through the Earth are not straight rays, but perhaps parabolas or something similar, his circles would still most certainly not meet at one point, even though such an analysis would still yield approximate information about the location of the earthquake. So how could Mohorovičić even know the exact distance of each instrument from the earthquake and thus determine the precise location of the earthquake, because it is evident that this is essential for any further serious analysis?

Maybe this will surprise people, but the key to success was in the parish offices. Namely, in the past, data on damage had been collected in schools, from doctors, and from parish offices because it was teachers, physicians, and priests who were literate and able to compile reports from the testimonies of their fellow residents. Then, the scientists and their assistants in their observatories would map the intensity of the ground shaking[10] based on the information collected on the damage in the field. From that map, the location of the earthquake's epicenter (the point on the surface of the Earth directly above the earthquake) would be determined, to put it simply, at the place of the most extensive damage. So, Mohorovičić was able to combine several sets of data and approaches to the problem and thus determine the location of the earthquake's focus and its unreliability to the extent that the conditions at the time allowed him to do so.

From the Mediterranean earthquakes that occurred a few years before the one in Pokupsko, he extracted data on how long P and S waves travel from the earthquake's focus to various seismometers on the Earth's surface. The farther the seismometer

10. See more about intensity in chapter 4.

is from the earthquake's focus—according to some rule—the more the S waves will lag behind the P waves, which can be shown by curves called *hodochrons*. This means that, although he did not know well what the fine details of the Earth and the path of the waves in the Earth's interior look like, based on his empirical hodochrons and assuming that the Earth's subsurface is more or less the same under the Mediterranean and southeastern and central Europe, he could use the triangulation method to obtain a result for the location of the Pokupsko earthquake, which was not bad at all.

However, the Pokupsko earthquake confirmed what had been bothering him for a long time. Instead of one P wave, he observed two of them on the large number of seismograms he collected! We expect one, and we get two! What did it mean, and how did it eventually lead to a great discovery?

———

"Hmm, would it work if I tune this a bit?" Andrija Mohorovičić asked himself, resolutely pressing another smoked cigarette in the ashtray, not taking his eyes off the neatly written lines of equations. He changed the initial assumption once more and started from the beginning. There were mathematical symbols ranging from trigonometry to vector calculus and numerous derivatives and integrals, which were probably still called infinitesimal calculus in his circles. The notes on his diagrams were, in addition to Croatian, also written in German and Latin. Although he taught in his mother tongue, he spoke English, Italian, and French fluently from his youth, mastered Latin and ancient Greek, and later added German and Czech to his list of languages! When you consider that the then schools of geophysics in Potsdam and Göttingen were famous world centers[11] because of their scientists, it is clear that the German language was ubiquitous in the early days of seismology. He could not help but recall the Hornstein, Mach, Durège, and Lippich lectures from his student days in Prague. Everything those experienced mathematicians and physicists taught him at the University of Prague came in handy for the problem in front of him on the desk.[12]

Instead of simple hodochrons—curves that show how long it takes waves to travel from an earthquake to seismometers at various distances—Mohorovičić was faced with the fact that the curves of P and S waves had two branches each. How can you unambiguously determine the earthquake's location from this, or, more importantly, how can you explain the phenomenon that on some instruments, there is only one, and on others, two P waves? Mohorovičić knew the physics of waves very well, and it was clear to him that the same laws of reflection, refraction, and diffraction (bending) must apply to seismic body waves or sounds as to light.[13] If you imagine that the entire spectrum of energy in which an earthquake emits its waves can be filtered so that only the high frequencies pass through, then the wave paths become rays, and it is possible to compare them with rays of light.

11. Emil Wiechert (previously mentioned as the scientist who proposed that the Earth has an iron core) was appointed the first chair in geophysics at Göttingen University, probably the world's first professor of geophysics. Beno Gutenberg was one of his students. In Potsdam, Ernst von Reuber-Paschwitz recorded the first teleseismic earthquake from Japan in 1889.

12. The fact that the department chair after the retirement of Lippich was taken over by Einstein for a couple of years (1911–1912) before he returned to ETH Zürich, his alma mater, speaks of how renowned the University of Prague was at that time.

13. Unlike a seismic wave, light is an electromagnetic wave.

The key difference between the speed of a ray of light and a ray of the P wave (or sound) is that light travels through a medium more slowly if you increase the density of the medium, while the P wave does the exact opposite—it moves faster and faster! It was already reliably known then that the density of the material increases from the surface to the center of the Earth. Namely, from the measurement of the density of the rocks available on the surface of the Earth and the estimation of the average density of the Earth,[14] it became clear that the density in the planet's center must be much higher than that on the surface. That is intuitive and understandable because, when you expose a material to such high pressures as prevail in the Earth's interior, the molecules and atoms of which it is made will be more compact than at atmospheric pressure. If there were no change in the density of the rocks, the wave rays would be straight. However, since P waves accelerate as they sink into the Earth's interior due to the increase in density, their paths will be curved outward in the direction of travel. The more pronounced the change in speed with depth, the more curved the paths will be.

To solve the problem before him, Mohorovičić had to assume that he knew how the speed of waves behaves at various depths in the Earth. In other words, he assumed that he knew the "model of the Earth," from which he then calculated the shape of the wave path using a direct approach. According to this model, the Earth is made of infinitely thin layers with a gradual change in density. From the first attempts, he failed to solve the problem because he did not immediately guess the best mathematical form that describes the behavior of the wave speed with depth. However, after several attempts, he managed

14. The average density of the Earth is obtained when its mass is divided by its volume.

to find a suitable expression that reduced the problem to only two unknowns: the speed of P waves at the surface, and the parameter of the increase in speed of P waves with depth. In other words, if the initial speed of P waves at the surface is known and the recipe according to which the speed increases with depth is also known, the speed of P waves at any depth in the Earth can be calculated accordingly.

When he inserted his initial assumption into the theoretical equations of the path of the wave and the travel time of the wave along that path, he managed to reduce the problem to a single mathematical equation, although, at first glance, it was rather complicated. In addition to the distance and travel time that he measured by analyzing the seismograms of his colleagues from all over Europe, two unknowns from the initial assumption were built into the equation. Generally speaking, you can solve a problem when you can express it as an equation with two unknowns and have enough measured data. That is precisely what Mohorovičić managed to do; he found a way to determine his two parameters and theoretically synthesize the observed data with great accuracy. That was a big step forward.

The problem of P and S waves appearing twice on some seismograms remained. Mohorovičić noticed that the second P wave disappeared about 760 kilometers from the earthquake's epicenter. The key was that he understood that this happens because the curved path of the P wave touches the layer beneath, which is much denser. This means that in addition to a gradual increase, we have a sudden increase in speed with depth—that is, a discontinuity. Why not, since, after all, it has been known since 1906 that the Earth has a core, and the discontinuity at the boundary between the core and the mantle is more pronounced than the very surface of the Earth where air and solid Earth meet! The jump in the speed of the waves at the discontinuity is large enough that the energy of the waves coming from above breaks into the lower layer and, at the same time, spreads along the boundary surface but cannot continue its journey in the upper layer along a curved path to the surface.

On the other hand, a discontinuity between the upper, slower and the lower, faster layers means that waves refracting at the discontinuity downward travel faster through the Earth than those in the upper layer. Seismographs at smaller distances thus record two P waves each, the slower of which travels through the upper layer and the faster through the lower layer. Seismographs at greater distances record only a single P wave surfacing steeply from the lower layer.

The discontinuity that Andrija Mohorovičić had to introduce into his Earth model to explain his observations was nothing less than the boundary between the Earth's crust and mantle. But what crust thickness did he determine, and what did this discovery mean for the scientific world? Is it possible to reach the Earth's mantle with today's technology?

In one—from today's perspective—unusually long article in the Annual Report of the Zagreb Meteorological Observatory from 1910, Andrija Mohorovičić presented the results of his analysis of the Pokupsko earthquake. In Croatian and German languages, he reported on how he discovered the discontinuity and pointed out many other directions and problems in seismology that developed into specialized fields in the following years. In his analysis, Mohorovičić showed that the horizontal discontinuity lies in the Earth at a depth of 54 kilometers. In other words, the crust in the area of the Earth for which he collected data is 54 kilometers thick. This depth agrees very well with modern measurements of the thickness of the crust for the northwestern part of Croatia and farther in a northwestern direction.[15] Namely, for a large number of data collected by Mohorovičić, the paths of the waves that started from Pokupsko toward European cities were sensitive to precisely that part of the Earth's underground.

In Mohorovičić's honor, the name *Moho* has been adopted for this discontinuity, both out of affection and due to the impossibility of pronouncing his last name in the wider world. Today, we know from modern measurements that the Earth's crust is globally present and that its thickness or the depth of the Moho varies. The Moho is thus shallowest under the ocean, in some places at a depth of only 10 kilometers or less. The deepest is under the great Himalayan massif, plunging 80 or more kilometers below the surface. This difference leads to the conclusion that it would be easiest to reach the Moho by piercing the oceanic crust.

We also talk about the Moho on the Moon and the other planets of the Solar System, which have differentiated into shells similar to Earth. However, one gets the impression that the average world knows very little about Mohorovičić's discovery of the boundary between the Earth's crust and mantle, at least from the perspective of a geophysicist.[16] Maybe this would be as shocking as the realization by a medical scientist that most of us don't know that penicillin was discovered

15. See the publication by Stipčević et al. (2011) in the bibliography.

16. I first heard about Mohorovičić's Discontinuity or Moho in elementary school and later in high school. During my university days at the Faculty of Natural Sciences and Mathematics in Zagreb, I was fortunate to learn the basics of seismology from Professor Skoko, who was a student of Mohorovičić's student, Dr. Mokrović, and also from professors Herak, Orlić, and other colleagues who came from the same school, established by Mohorovičić himself. I am infinitely grateful for their insights.

by Alexander Fleming in 1928—except perhaps in Japan, the country in which Mohorovičić is well known due to the presence of many seismologists and Earth scientists.

If you speak with the locals of Volosko in the local pizzeria Moho, they remember their famous neighbor, a giant of the geophysics world, mostly only when the ground trembles under their feet, and not being able to explain very well what is so important and how their fellow neighbor worked it out and thus indebted the world scientific community. Quite the opposite of a taxi driver in Alaska, whom I encouraged, due to his enthusiasm, during the ride from Fairbanks airport to explain the physical mechanism of the aurora borealis to me, and he did a great job! Who knows, maybe he was a physicist from the former USSR!

In Vinkovci, among the oldest continuously inhabited places in Europe, on the edge of the Pannonian Plain on Roman foundations, as a child, I once found a coin with the head of a Roman emperor. By coincidence, Mohorovičić also ended up on a gold coin one day. It's a remarkable turnaround in human mentality that instead of military leaders and politicians, we start putting scientists on national commemorative coins! Today, the Moho is on postage stamps, in the names of lunar craters and asteroids, in science fiction stories, and in video games.

Anyway, recalling the discussion about direct and inverse methods in science, Moho's approach can be described as a combination of both. In fact, without a well-defined direct problem, there is no solution to the inverse. In this story, his meticulously made observations of two P and S waves, each at a specific interval of distance from the earthquake, and, not least, the openness of his intellect, were crucial. This openness to always favor observations to theory led him to discovery. Therefore,

Moho's method has the refined taste of a solved inverse geophysical problem very early in the history of seismology.

––––––

Andrija Mohorovičić died in 1936, symbolically, the same year that another great discovery saw the light of day—that of Inge Lehmann, that the Earth has an inner and outer core. The discovery of the boundary between the outer and inner cores some 26 years after Mohorovičić's article is kind of equivalent to the discovery of the boundary between the crust and the mantle.[17] So, let's dwell briefly on Lehmann's discovery and compare it with Mohorovičić's.

When Inge Lehmann, a shy Danish seismologist in her fifth decade of life, analyzed records of earthquakes from faraway New Zealand in her office at the University of Copenhagen, roughly on the other side of the world from those earthquakes (see the path of seismic waves that traverse the Earth's core in figure 1.1), she could not have known that her work on them would celebrate her. She worked on the earthquakes near Buller in 1929 and Hawke Bay in 1931. Since the earthquakes were on the antipodal side of the globe from Europe, the paths of the compressional or P waves passed almost through the very center of the Earth. At that time, the existence of the liquid core of the Earth and the boundary between the core and the mantle approximately halfway between the surface and the center of the Earth was well documented by Beno Gutenberg and his

17. Both discoveries are very dear and close stories to me because Mohorovičić's procedure convinced me to choose solid Earth physics as my specialty and continue my geophysics doctoral research. The inner core of the Earth was the subject of my doctoral dissertation and is still part of my daily research.

seismological colleagues and accepted as fact. At large distances from the epicenter, which was the case with the distance of the European stations from New Zealand, P waves that pass through the center of the Earth, as well as those that reach much shallower depths in the liquid core, were visible. However, Lehmann noted no satisfactory explanation for another branch on the observed hodochrons. This was reminiscent of Mohorovičić's observations, only a much deeper phenomenon. One possible solution was waves that would oscillate around the core-mantle boundary, but they were too weak to explain Inge Lehmann's observations.

Another possible solution to the problem, perhaps the most elegant of all, was that the Earth has another core within the liquid core, manifesting as a discontinuous increase in the material density and velocity of seismic waves. We can return for a moment to the discussion of the inverse and discrete approaches and imagine the then model of the Earth's internal structure as a model with another sphere in the center of its core. It's as if we discovered another *matryoshka* or *babushka*, a traditional Russian figure made of linden wood in a set of nesting dolls that hide even smaller copies inside. This new model could accurately predict the occurrence of three types of waves and did not contradict other observations. Lehmann herself was cautious, indicating in her famous article, the shortest name in the history of seismology: P'[18], that this model would survive only if it could confirm future observations. Indeed, future observations confirmed her model even though the inner core as we know it today has a slightly smaller radius than

---

18. See the publication by Lehmann (1936) in the bibliography. P' waves are also known as PKP waves (see the captions to figure 1.1).

that initially determined by Inge Lehmann[19] (fig. 1.1). It should be noted that in this model, the inner core of the sphere is without a clearly defined aggregate state.

A few years later, when the Second World War was raging in Europe, the American mineral physicist Francis Birch hypothesized that the discontinuity that Lehmann observed could be the transition from a liquid to a solid aggregate state of iron at high pressures. Although several previous works reported the observation of shear waves in the inner core, we only recently put the "dot on the i" with empirical confirmation that shear waves also propagate through the inner core, which is, consequently, in a solid aggregate state.[20] Interestingly, we extracted this from the correlation wavefield, using the method by which, instead of the most prominent earthquake signals, we used records of only weak ground motions hours after the occurrence of large earthquakes in the so-called earthquake coda. This result also marked the end of the 80-year search for shear waves in the inner core—evidence that it is in a solid state because shear waves cannot move through liquids. The search for these waves was often referred to in seismological circles as the search for the holy grail of observational seismology.[21]

19. She determined that the radius of the inner core is about 1,400 kilometers, while today's accepted value of the radius is about 1,220, with an uncertainty of several kilometers. See, for example, the publications by Dziewoński and Anderson (1981) and Kennett et al. (1995).

20. This study is the contribution of my research group at the Australian National University. See the publication by Tkalčić and Phạm (2018) in the bibliography.

21. The title of section 4.8.3 of the textbook *Introduction to Seismology* by Peter Shearer (2009) is "PKJKP: The Holy Grail of Body Wave Seismology." I had the honor to reflect on Inge Lehmann and her discovery of the Earth's inner core in a video by the American Geophysical Union, released on the occasion of the centennial of its founding. See the video link in the bibliography.

I often wonder what Inge Lehmann would say about confirming shear waves' passage through the inner core. Or what would she say about the relatively new finding that the inner core has another, smaller core inside it?[22] What would Andrija Mohorovičić think about everything we have discovered about the Earth's interior since his time? I would like to sit down with him for a coffee somewhere in Grič,[23] show him the confirmation of the global existence of Moho on my laptop, a map of the thickness of the Earth's crust, today's seismometers including those that go to the ocean floor and the surfaces of other planets, discoveries about the Earth's core, mantle and crust, as well as advances in the understanding of earthquake physics and engineering seismology. I keep thinking that the coffees would be lined up one after the other.

Today, when, as a human species, we think about how much we know about the interior of the planet on whose surface we share this moment in the infinity of time, we realize that we have climbed onto the shoulders of Mohorovičić and Lehmann, giants of geophysics, and stand firmly on them. New knowledge is born on the horizon. The missions of the fantastic research drilling vessels *Chikyū*[24] and *Meng Xiang*[25] to pierce the

22. The innermost inner core was first proposed by my colleague, Miaki Ishii and her PhD supervisor, the late Adam Dziewoński, in 2002. See the publication by Ishii and Dziewoński (2002) in the bibliography.

23. Medieval settlement on a hill; a part of Zagreb's old town, where a monument was erected to honor and remember Mohorovičić (figure 3.4).

24. From the Japanese word for Planet Earth. This scientific research vessel was designed and built to pierce the thin oceanic crust and reach the Earth's mantle. Its gross tonnage is 57,000 tons, and its length is 210 meters.

25. From the Chinese word for Dream. Its gross tonnage is 33,512 tons, and its length is 180 meters. The ship is designed to drill through the Earth's crust and into the upper mantle, contributing to the global efforts to explore the Earth's interior. Its trial voyage was completed in December 2023.

FIGURE 3.4. The monument to Andrija Mohorovičić is erected in Grič, the old town of Zagreb, Croatia, near the Meteorological Service building where Mohorovičić worked. The sculptor, Nikola Džaja, designed it symbolically without a base, showing Mohorovičić rising from the Earth's crust.

Earth's crust, reach for the first time the Mohorovičić Discontinuity, the Earth's mantle, and collect rock samples from those depths, are the symbols of all our dreams, victories and struggles, aspirations, and hopes for a better future.

The illuminated path is the legacy of great scientists who, not so long ago, turned their gazes from the sky's direction toward the Earth's depths and unveiled its secrets.

4

Giants that sometimes wake up

Putting the finishing touches on the text of his article published in 1935 and rubbing his palms together with satisfaction, Charles Richter looked up at the starry sky of Southern California and thought what a great job he had done in introducing the concept of magnitude into seismology, from astronomy—his great childhood love. Back in Hellenistic times, Ptolemy used six magnitudes as a measure of stars that could be seen with the naked eye. But in modern astronomy, magnitude is used in the context of the relationship between distance and the actual luminosity (brightness) of stars or galaxies, which is the amount of energy they emit per second. A higher luminosity means the star is brighter; that is, it produces and radiates more energy per second than the other star. However, the stars are at different distances from us, and the brightness intensity decreases with the square of the distance; for example, 4 times the distance increase results in a 16 times smaller brightness. The naked eye perceives some of them as less bright, which may be because they are much farther away from us.

That is why it is necessary to introduce the concept of absolute magnitude, a measure that refers to real instead of "apparent" luminosity. An object's absolute magnitude is the magnitude of its luminosity, which we would measure under ideal conditions if we placed all objects at the same distance from us at precisely 10 parsecs.[1] This will result in significant differences in actual luminosity among stars because the universe is vast, and the variation in the characteristics of space objects radiating energy is enormous.

We don't even have to go into space to figure it out. We can, for example, compare the size of a grain of sand from the white beach of a tropical island with the size of a big mountain. Let's take a more prominent grain of sand, about 0.8 millimeters in diameter, and compare it to the height of the Himalayas, which is about 8 kilometers. One kilometer is 1 million millimeters, so we would need about 10 million grains of sand stacked in a row to reach the height of the Himalayas. It's the same with luminosity. Our galaxy, the Milky Way, has a luminosity about 10 million times greater than our Sun. Those are huge and unwieldy numbers, but we can get to them by introducing logarithms. Let's recall, the logarithm of 10 is 1, the logarithm of 100 is 2 . . . the logarithm of a million is 6, and the logarithm of 10 million is 7! In other words, the luminosity of our galaxy is seven orders of magnitude greater than the luminosity of the Sun. The beauty of logarithms lies precisely in the simplicity of counting zeros. And yes, the magnitude can be negative: the logarithm of 0.1 is −1, the logarithm of 0.01 is −2, etc.

In the 1930s, while Inge Lehmann was studying the New Zealand earthquakes, Richter was working on a doctorate in theoretical physics at Caltech. He got his first job at the Carnegie Institution in Washington before completing his doctorate.

---

1. A measure of distance in astronomy. It is 32.6 light-years or about 31 trillion kilometers.

He became interested in seismology, and that's when he began his collaboration with Beno Gutenberg, which would eventually bring world fame to Caltech's seismological laboratory.

According to the relation established by Richter in collaboration with Gutenberg, the earthquake's magnitude is determined using logarithms so that it is calibrated at a distance of 100 kilometers. Thus, a magnitude of 3.0 corresponds to an earthquake at a distance of 100 kilometers, causing a maximum movement of 1 micrometer[2] on the seismograph used at Caltech at the time. Suppose a larger earthquake caused a maximum movement (displacement) of 1 millimeter under the same conditions.[3] An increase of 1,000 or three orders of magnitude means that the magnitude of that earthquake equals 6.0. Over time, the name local or Richter magnitude became established for this method of magnitude determination.

However, neither Richter nor Gutenberg could have imagined how much confusion their original approach to earthquake magnitude would create in recent years among the citizens of other world cities who only occasionally experience shaking. Of course, the problem is that the Richter magnitude calculation works quite well for Southern California, for relatively close earthquakes, which are always recorded on the same type of instrument. But today, we know that the intensity of shaking[4] depends not only on the amount of energy released

2. A thousandth of a millimeter.

3. A distance of 100 kilometers and the same seismograph.

4. Unlike earthquake magnitude, the intensity scale is based on the earthquake's effect on objects on the Earth's surface. An example of an intensity scale is the Mercalli-Cancani-Sieberg scale, where the intensity varies from I to XII. On that scale, an earthquake of intensity I can be detected only by seismographs, and an earthquake of intensity XII is a catastrophic earthquake that destroys all human-built infrastructures and permanently changes the relief of the Earth.

by an earthquake but also on the physical mechanism of the earthquake, on the internal structure of the Earth through which seismic waves travel, on the type of soil on which the building containing the seismograph is located, and several other considerations. Therefore, over time, a new measure was introduced for earthquake magnitude, called the magnitude of the "moment" or simplified, *moment magnitude*. According to it, the absolute magnitude of an earthquake depends only on the amount of energy released. Put simply—take a deep breath—the difference in magnitude between two earthquakes, M2 and M1, is equal to the product of 2/3 and the logarithm of the ratio of their seismic moments, MO2 and MO1.[5]

Mathematically, it can be written like this:

$$M2 - M1 = 2/3 \times \log(MO2/MO1).$$

The seismic moment, MO, which is related to the energy released by earthquakes, is obtained by multiplying the strength of the rocks, the activated surface of the fault, and the average displacement on that fault. This requires knowing the parameters that can be calculated by seismological methods or observations on the Earth's surface. The above relation eliminates the need to read the displacements from a specific seismograph and calibrate them by distance via a method suggested by Richter and Gutenberg. There is no displacement or distance in the formula. So, if you ever run into a question about what it means for an earthquake to be ten times "stronger" than another (stronger in the sense that more energy is released), or what the difference in magnitude would be if one earthquake were 32 times stronger than the other, and so on, feel free to come back

---

5. For more information on the seismic moment, see the glossary of relevant terms.

here for the answer. Take, for example, a moderately strong earthquake of magnitude M1 = 5.3.

Using the above relation, let's first look at the magnitude M2 of an earthquake ten times stronger than that.

So we have the following:

$$M1 = 5.3; M2 = ?$$

Ten times more energy means: $10 \times MO1 = MO2$ or, if expressed as a ratio: $MO2 / MO1 = 10$.

What is the magnitude of M2? From the above equation for the difference in magnitudes, it then follows:

$$M2 - 5.3 = 2/3 \times \log(10).$$

The logarithm of 10 equals 1, so we have only 2/3 remaining on the right side of the equation.

$M2 - 5.3 = 2/3$, and therefore:

$$M2 = 2/3 + 5.3 = 0.67 + 5.3 = 5.97.$$

It follows that the magnitude of an earthquake 10 times stronger than that of magnitude 5.4 is approximately equal to 6.0.

As another example, let's now ask how many times an earthquake of magnitude 7.0 is stronger than one of magnitude 6.0.

It follows from the above formula:

$$7.0 - 6.0 = 2/3 \times \log(MO2/MO1)$$
$$1.0 = 2/3 \times \log(MO2/MO1)$$
$$3/2 = \log(MO2/MO1)$$

$MO2/MO1 = 10^{3/2} = 31.623$, which is approximately equal to 32.

Therefore, a difference of 1.0 in the magnitudes of two earthquakes means that the second earthquake released about 32 times more energy than the first. A magnitude difference between two earthquakes of 2.0 means that the second earthquake released exactly 1,000 times (31.623 × 31.623) more energy than the first. A magnitude 7.0 earthquake will release 1,000 times more energy than a magnitude 5.0 earthquake, but we should also say that this energy will be released for a longer time and across a larger surface than magnitude 5.0. Now that you are armed with the knowledge of logarithms and magnitudes, things should be a little clearer.

———

In the early morning hours of 22 March 2020, an earthquake with a magnitude of 5.3 beneath Medvednica Mountain,[6] a few kilometers north of the city, shook Zagreb and its wider surroundings (fig. 3.1 shows the fault lines north of Zagreb). It is difficult to say how many were surprised, but it caught even some seismologists mentally unprepared, perhaps mainly because it happened in the immediate vicinity of the city of Zagreb. It caused material damage to the city and its surroundings. The good thing about that event is that it was a relatively moderate to moderately strong earthquake. If by any chance it had been a slightly larger earthquake, say magnitude 5.7—as was the case the day before on the border of Greece and Albania—or an earthquake of magnitude greater than 6.0, we could use the above examples to calculate that the released energy would be tens of times higher. Therefore, with all due respect to all the victims, it must be said that it could have been a lot worse, but fortunately, it wasn't.

---

6. The mountain is named after the bears that once lived on it. Traces of Neanderthal life from 45,000 years ago were found in the Veternica cave.

In the days immediately after the earthquake, the citizens of Zagreb and the surrounding area were most interested in what to do and what to expect. Some began to doubt their decision to move here from provincial towns, while others wondered whether they should stay in their homes or temporarily leave the city and seek salvation somewhere in the Slavonian plain or on the Adriatic coast. After initially questioning the purpose of the seismology profession, most people slowly accepted the critical difference between meteorology and seismology. They realized that the faults on which earthquakes occur deep beneath our feet are inaccessible to direct examination. They also slowly accepted that earthquake forecasting is not the same as weather forecasting and that no one could and cannot predict future earthquakes with certainty. They realized that seismologists don't know exactly how big the faults are; they don't know the geometry of the faults in the Earth's interior very well or why the rock, once it breaks and ruptures in one place, stops rupturing in another place. It was slowly accepted that this invisible underground architecture should first be "imaged" using modern seismological methods, just as radiologists use tomography to image the inside of the human body.[7]

Although they could not predict future earthquakes, seismologists knew from previous examples that there was a period of "healing" of the fault where the earthquake occurred, which means that the citizens of Zagreb and the surrounding area expected increased seismic activity and many moderate and weak earthquakes. Indeed, the largest aftershock had a moment magnitude of 5.0. One of the first things seismologists did was to go out into the field and try to find a trace of the fault where the earthquake occurred on the very surface of the Earth. After

7. We will discuss imaging in chapter 6.

shallower earthquakes, you can sometimes see a fault line (at the intersection between the fault plane and the Earth's surface) along which two Earth blocks moved relative to each other and thus caused the earthquake.

When an earthquake is shallow, shear along the plane usually breaks the Earth's surface. Manifestations of faults on the Earth's surface help determine the fault on which the earthquake occurred, because mathematical solutions—due to the very nature of shearing and forces in the quake's focus—are ambiguous. In other words, from seismogram analysis, two mathematical solutions, that is, two planar surfaces, are obtained by computer programs. These two planes are perpendicular to each other, and we can't say—by non-uniqueness—on which one of them the slip occurred. Visualizing faults on the Earth's surface can offer a unique, physical solution and tell us a lot about the orientation of the fault surface itself: how the two rock masses in contact moved, in which direction the most energy was released and, consequently, where the most structural damage and possible victims may be. However, the problem arises when there are no traces of faults on the surface.

Fortunately, from the distribution of locations of smaller earthquakes, the extent of the fault zone that was activated can be subsequently determined, and whether it is a brand-new fault that we did not know about yet or a known fault, but these investigations will last for years. In fact, the aftershocks are something the inhabitants of California and Japan are entirely used to. Because of earthquake preparedness, people in many places have learned to live with earthquakes. However, it is always necessary to correctly identify the damage and determine whether it is possible to continue living in the damaged house or building.

In Zagreb, installations should have been thoroughly checked and possibly closed after the earthquake. In the immediate aftermath of the quake, many people were thinking about where it would be best to go or, if they stayed, where in the house or apartment was the safest. Therefore, the message had to be conveyed to the population that if they felt an earthquake, they should not run away from buildings or houses while the ground was still shaking—the most common advice was to seek protection and cover their head, even under a table.

The earthquake under Medvednica was not caused by two tectonic plates as in San Andreas but by two blocks of rock in contact: its location is quite far from the edges of the tectonic plates. But the very existence of Medvednica is proof that the area is geologically active. During the Earth's geological history, some parts of its crust rose, others descended, and Medvednica was once an island in the vast Pannonian Sea.[8] Geologists know there are faults in the area of Medvednica; after all, the great earthquake of 1880 happened on one of them.[9]

Material damage depends not only on the direction in which the most energy was released and how far the location is from the earthquake's epicenter but also on the type of soil on which a building, house, school, church, road, or bridge is built. It is undoubtedly very difficult to produce a detailed map of the underground through which the seismic waves spread. The structure of the Earth determines how much of the initial energy released from the quake's focal point or a ruptured planar fault surface will reach a particular location on the Earth's surface.

The initial seismological analyses showed that it was reverse faulting, where an earthquake occurs because two blocks in

8. The Pannonian Sea existed from about 10 million BC, when the rise of the Carpathians separated it from the prehistoric sea Paratethys, until about 1 million BC, or according to some sources, 600,000 BC.

9. The great Zagreb earthquake of 1880, whose magnitude is estimated at 6.3.

contact have been subjected to compression for a long time, just as in the packing-sponge example. Let's recall that at the San Andreas Fault, which runs along the American coast almost perpendicular to the surface of the Earth, rock masses move in the direction of the fault. Reverse faults have a dip relative to the Earth's surface, and the rock masses move perpendicular to the direction of the fault in a fashion that the upper block moves up relative to the lower one. They are the opposite of normal faults, in which the blocks are exposed to dilatational forces, and the upper block moves down relative to the lower one.

Because of the way the ground moves in P waves, these motions feel to humans like either a compression shock or a dilation collapse. They can throw you out of bed, or you can feel like you lost the ground under your feet, but they rarely cause significant damage. That is precisely why S (or shear) waves, in which ground particles move perpendicular to the direction of wave propagation, caused the most significant damage in Zagreb. This type of motion is analogous to what most people describe as a "rocking" sensation. The rocking in moderate earthquakes can last for a few seconds and, in stronger ones, for about a minute or longer. At greater distances from the quake's epicenter, it is not P or S but surface waves that cause the most damage. We will return to this point later.

After the earthquake gave them a good shake, people's patience and nerves were put to the test by the restless Earth for weeks and months with subsequent tremors. It was as if someone had thrust them into a large natural laboratory without warning and forced them to sleep in their clothes and shoes. They were repeatedly told that this could take weeks, months, or even longer after the main event based on studying earthquakes in other tectonic conditions in different parts of the world. They were also told that minor aftershocks occur near the main quake's epicenter, mainly on the fault plane beneath the Earth where the main shock occurred or on adjacent fault planes near the epicenter. That is because the underground rock architecture "gets used to" the new conditions with new redistribution of pressure in the Earth's interior after the main shock, if we can describe it that way.[10]

Many have rightly wondered what the maximum earthquake magnitude is that could be expected in the future. The most popular scientific method for determining the maximum magnitude in a place consists of a statistical model whose basic assumption is that the number of earthquakes of a certain magnitude in a specific area and during a particular time can be determined according to the well-known Gutenberg-Richter law, the two scientists we have already mentioned. Although the law is quite universal—in a given population of earthquakes, most will be weak, and there will only be a few strong ones—there are variations from area to area or fault to fault. The key is to have an excellent catalog of earthquakes, including those below human perception, which can be felt only by sensitive instruments. That is why seismological services and agencies that monitor the seismicity of an area or country must be well-equipped with seismographs.

There can be hundreds or even thousands of instrumentally recorded aftershocks, but their frequency decreases with time, according to a law known as Omori's law.[11] It says that the frequency of earthquakes decreases inversely proportional to the time that passes after the main shock. This, in a simplified way, means that the chance of an aftershock occurring on the second

10. And why there are forces and pressures in the Earth's interior was described in the introductory chapter when we talked about plate tectonics.

11. According to a seismology pioneer from Japan, Fusakichi Omori.

day will be two times less than on the first day, and on the tenth day 10 times less than on the first day. This relationship was arrived at empirically by observing a large number of earthquakes. Actual data for individual earthquakes behave stochastically, meaning they also have elements of unpredictability but mostly follow Omori's law. Determining the locations and times of the aftershocks is vital because they can be used to characterize the geometry and size of the fault plane where the sudden slip occurred during the main shock.

In addition to P and S waves, which we can use to locate earthquakes via a triangulation method, we can learn much more from ground motion records after an earthquake. Let's take two seismograms as an example: the first, recorded in the vicinity of the Zagreb earthquake, on Puntijarka, only 12 kilometers from the epicenter as the crow flies, and the second, in Morići, on the southern side of Lake Vrana near Šibenik, about 235 kilometers from the epicenter as the crow flies (fig. 4.1; for the locations see fig. 3.1). Apart from the time series records of ground motion over time (seismograms), let's look at another type of data from those two seismological stations, data which exist because seismic waves are not monochromatic.[12] Namely, the energy from the earthquake's focus was released in a broad spectrum of frequencies. That is why—in addition to seismograms—it is also possible to see how strong the signal is recorded in various parts of the frequency spectrum over time in a record known as a spectrogram. While the seismogram looks like a printed curve in time, the spectrogram resembles a colored bar (fig. 4.1).

Let's focus first on the five-minute seismogram and the corresponding spectrogram from Puntijarka, with the same time

---

12. In other words, ground particles do not vibrate at just one frequency.

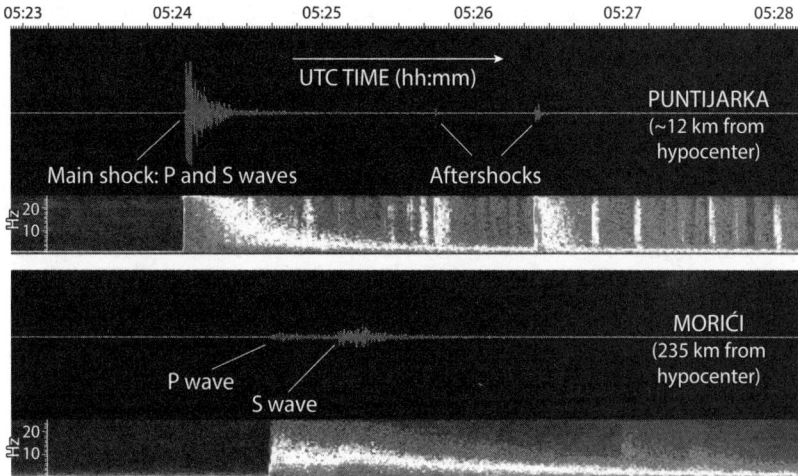

FIGURE 4.1. Seismograms and spectrograms from the 2020 Zagreb earthquake, recorded at the seismological stations Puntijarka (top two diagrams) and Morići (bottom two diagrams). The horizontal axis of the seismogram shows time, and the vertical axis shows the speed of moving ground particles during the passage of seismic waves. The horizontal axis of the spectrogram shows time, and the vertical axis shows the frequency of ground particle oscillations. The colors show the strength of the frequencies—stronger frequencies are shown in red, and weaker frequencies are shown in blue. At the Puntijarka station near Zagreb (fig. 3.1), P and S waves arrive almost simultaneously, and at the Morići station near Šibenik (fig. 3.1), the arrivals of P and S waves are visibly separated. The original seismograms and spectrograms were graphically presented by Marija Mustać Brčić from the Seismological Service using the "Scream" software (see the page: https://www.guralp.com/sw/scream.shtml.)

scale on the horizontal axis. We can extract a seismogram from continuous data that show the speed of movement of ground particles in a time interval of about one minute before and four minutes after the earthquake's origin time. The corresponding spectrogram shows us how the strength of the wave frequencies changed in that same time interval. The color spectrum represents the frequency spectrum; the red part indicates higher, and the blue has lower energy.[13] In simple terms, frequency refers to how many oscillation cycles one ground particle went through per second during the passage of seismic waves; that is, how many wave crests or troughs changed in one place in one second. For example, a frequency of 10 hertz means that a ground particle vibrates 10 times in one second. This information may be helpful when we touch on rumble and sound effects.

The P waves that traveled the fastest through the Earth from the earthquake's hypocenter were recorded at Puntijarka at 6 hours, 24 minutes, and 3 seconds at local time (note that figure 4.1 shows the universal time). Perhaps only about a second after P waves arrived, the S waves also arrived—slower than P waves but with much larger amplitudes. The time gap between the P and S waves was tiny due to the proximity of the Puntijarka seismometer to the earthquake's hypocenter. Let's remember the example of lightning and thunder again and how we determined the distance to the source. Due to the seismometer's proximity to the earthquake's focus, we cannot expect to see the P and S waves separated in time, but from the seismogram, we can effortlessly conclude that the focus is very close. We can see the arrival of seismic waves very well on the spectrogram, which shows with a broad red band that a large part of the energy of the waves was present in the interval between 0 and 25 hertz for a good 20 or more seconds. After that, the energy gradually decreases at higher frequencies and survives only at the lowest ones. The red strip thins out, and after the passage of body and surface waves, it gradually sinks back into the blue of the ambient noise.

13. We will return to this when discussing quakes on Mars.

And what can we see on the seismogram and spectrogram in Morići, much farther away? The critical difference between this and the seismogram at Puntijarka is that the arrivals of P and S waves are separated and visible here. Hence, we can see that the S waves arrive about 26 seconds after the P waves. If the average speed of P waves through the Earth were 6 kilometers per second, and the average speed of S waves about 3.6 kilometers per second,[14] this converts to a travel time of 235/6 = 39.1 seconds for P waves and 235/3.6 = 65.2 seconds for S waves for a distance of 235 kilometers. That corresponds to a time difference of about 26 seconds; we can notice this time difference on the seismogram with ease.

It is insightful to look at the time before the appearance of P waves on the spectrogram at Morići. In addition to the dark blue color that prevails at all frequencies, we can notice the ambient noise in the light blue right at the bottom for frequencies approaching zero. As it was early morning, we assume that a large part of that ambient noise could be attributed to the proximity of the Adriatic Sea and Lake Vrana. Amplified frequencies can also be present due to the meteorological conditions that prevailed that morning. Further, we can see that the red color predominates at the beginning only at frequencies up to about 5 to 6 hertz. Later, the red color is only present at lower frequencies. The primary difference between this and the spectrogram on Puntijarka is that the energy at higher frequencies was lost when the waves passed through the Earth's subsurface from the earthquake's hypocenter to Morići.

But let's now return to Puntijarka and focus on several aftershocks visible on the seismogram. As you now have a certain degree of knowledge about the spectrogram, you could very

14. The speed of S waves is usually about 60% of the speed of P waves.

quickly notice aftershocks on the spectrogram as well. They are much less intense and last less time. You may also see a series of other aftershocks recorded by this instrument, all within the first four minutes of the main quake. On the way to Morići, most of the waves of these aftershocks lost their energy so that their traces on the seismogram and spectrogram of this instrument are hardly seen there. In practice, if we had a high density of seismographs in Zagreb and its surroundings, the analysis of aftershocks would have been much simpler and more precise. That would result in a better understanding of the main shock and also a better understanding of the fault system. As this was not the case, seismologists had to rely on records from distant seismographs where high frequencies were lost, and aftershocks were often unrecorded. Due to this fact, after the earthquake, it was justified to hope that financial resources would be invested in the Seismological Service, for perhaps a few more instruments, but also basic research into the physics of earthquakes and the structure of the Earth beneath the whole of Croatia.[15]

———

In the days after the event, people witnessed the aftershocks that followed the main shock in large numbers. However, in addition to the trauma experienced, everyday psychosis was intensified by phenomena such as the feeling of slight shaking, the sound of rumbling, thunder, and distant explosions. Although sounds do not cause harm, they can contribute to fear and psychosis and should, therefore, be sought to be explained.

15. One such contract, worth €2.1 million, was signed in 2022 and is implemented through the financial mechanisms of the European Economic Area, the Norwegian State Fund, and the State Fund of the Republic of Croatia.

This phenomenon intrigued me long ago, partly because I experienced it once during my graduate student days in my tiny studio in Berkeley.

During an earthquake, energy is released in the form of body waves that travel through the Earth from the focal point to a seismological station, where a seismograph records them as a time series of movements of the ground where it is placed. It may sound trivial to mention that the ground particles do not move together with the wave from the focal point to the surface of the Earth where the seismograph is located. In this context, the movement of ground particles simply means that when a wave disturbance passes through a specific volume of the Earth, the particles at that place oscillate. Let's recall that oscillation in space can be described as an oscillation in the same direction as the wave passes (P waves) or is perpendicular to it (S waves).

During the P wave, energy is transferred from particle to particle by compression (contraction) or dilation (expansion). That's why you will either feel a sudden blow from below or a sudden pull down when you experience a P wave. On the other hand, when the S wave comes in, a rocking is felt instead of a shock or collapse. To help you visualize it, imagine a hula dancer moving in one direction, swaying their hips perpendicular to the direction of movement. Rocking caused by S waves during strong earthquakes has a destructive effect on buildings and infrastructure.

When an earthquake occurs, say at a depth of 10 kilometers, and the seismometer is directly above, the first to reach the seismometer will be P waves that arrive from beneath the Earth's surface along a vertical path. But suppose the seismometer is quite far from the epicenter. In that case, the waves that reach first from the focal point will sink deeper into the Earth's interior, where their propagation speed is higher. They will bend and then follow an almost vertical path to the seismometer on the Earth's surface. In our previous example, the P waves that first arrive from the focal point under Medvednica to Morići will sink into the lower layers of the Earth's crust, to a depth of about 35 kilometers, and then follow an almost vertical path toward the surface to Morići. But what if the seismometer is farther away, even on the other side of the world? For the P waves of a strong earthquake from the Mediterranean to be observed, for example, in Canberra, Australia, they need to travel to great depths, in the case of Zagreb and Canberra, even to a depth of 5,150 kilometers, which is in the inner core of the Earth, and then out through the outer core and mantle almost vertically from the bottom to the surface. That trip through the Earth's center would take under 20 minutes.[16]

With a P wave, ground particles oscillate in the direction of wave motion, similar to the way air molecules oscillate in the direction of sound propagation. In fact, by their very nature, P waves are the same as sound waves! It is certain that at least some of you have experienced sounds underground, if not in a mine, then in the subway of some distant city. From these examples, you know that sound propagating through the Earth can pass into the air and thus continue its propagation. It also spreads through water more efficiently than through air, which you must have noticed if you have ever dived under the water's surface somewhere at sea or in your bathtub.

For the same reason, it is not unusual that P waves that reach the Earth's surface from an earthquake can continue traveling through the air. P waves also propagate through the liquid outer core, unlike S waves, which cannot. Thus, the sounds associated

16. Remember that Inge Lehmann, studying earthquakes from New Zealand in her observatory in Denmark, discovered the existence of the Earth's inner core!

with earthquakes are produced above the Earth's surface when P waves from the Earth at the free interface of the solid Earth and the atmosphere or ocean are partially converted into sound waves. One of the most recent examples is the eruption of the Hunga Tonga–Hunga Ha'apai volcano in the South Pacific. The sound was heard, for example, in New Zealand and Alaska, thousands of kilometers away.

But the matter is not quite so simple. Namely, the human ear is sensitive to frequencies from about 20 to 20,000 hertz, while the P waves of earthquakes dominate mostly at frequencies between 0 and 5, often up to 10 hertz. In other words, the frequencies of P waves are in the so-called infrasound part of the spectrum, below frequencies of 20 hertz, to which the human ear has no sensitivity. So, one rightly wonders if the sounds people have reported for the main shock and aftershocks are even related to P waves. Look at the definition of infrasound. You will see that, apart from the fact that everything under 20 hertz is considered infrasound; earthquakes, volcanoes, landslides, avalanches, and meteorites are mentioned as natural sources of infrasound. Admittedly, elephants and hippos communicate on infrasound frequencies—so to speak, on their private communication channel—and it is not unusual for some people to have a slightly narrowed or widened range of frequencies they can hear. So, we should go back to the spectrogram and study all the frequencies observed on it a little better, with the remark that it can also be called a sonogram or infrasonogram if we deal with it exclusively to study sound.

The appearance of the spectrogram—a visual display of the relative strengths of frequencies over time—will be influenced by many factors, of which at least two should be mentioned: (a) the character of the earthquake itself, and (b) the structure of the Earth through which the waves propagate. As for the character of earthquakes, we can draw a perfect parallel between them and musical instruments. Larger instruments will generally produce a broader range of sounds, more vibration, and a greater range of frequencies, including even those above 10 hertz. That was the case with the Zagreb earthquake. Remember that the red color on the spectrogram reached very high frequencies, even up to 25 hertz. Furthermore, due to the Earth's structure, energy loss occurs when the waves pass through it because part of the energy is simply absorbed on the atomic scale through the vibrations of atoms and molecules. Higher frequencies weaken faster than the lower ones so that, eventually, there are no longer any high frequencies at a certain distance from the earthquake.[17] Let's recall that we observed this when we compared the spectrograms on Puntijarka and Morići.

It is precisely these high frequencies during an earthquake, which are also the lowest frequencies to which the human ear is sensitive, that are responsible for the sound that accompanies the main shock and probably also for a large part of the sound effects during subsequent earthquakes that people continued to witness en masse, especially near the epicenter. In other words, at the frequencies at which the ground particles vibrate due to the energy of seismic waves, the Earth's surface oscillates like the diaphragm of a large speaker and transmits sound into the atmosphere. In some cases, the sound may be amplified due to the topography of the terrain, for example, the presence of mountains. If the earthquake is relatively shallow and almost directly beneath you, you might hear it even without feeling any vibrations or rocking. However, sound does not travel very far from small or deeper earthquakes at these frequencies bordering on

17. This phenomenon is called attenuation, and we will mention it in connection with the propagation of body and surface waves and more in chapter 6.

sound and infrasound, meaning that even seismographs at greater distances would not record it, as some reports have suggested.

Let's also look at the results of two studies: one based on 77,000 reports of shallow earthquakes in Italy[18] and another on thousands of reports of earthquakes from the French Pyrenees[19] in which the percentage of people who heard the sound in a specific location was analyzed. The results of these two studies are very similar: the main conclusion is that the frequency of sound observations during smaller earthquakes is inversely proportional to the distance from the focal point, the depth, and the magnitude of the earthquake. How effectively sound spreads through the air depends on the local geology and terrain configuration, as well as on the current state of the atmosphere, for example, pressure, temperature, and wind. The influence of noise due to human activities in populated areas should not be excluded either. The limit of hearing is the line corresponding to the amplitude of motion of ground particles of 0.03 micrometers due to P waves. This means that earthquakes of magnitude 2.0 and above should be heard within a distance of about 10 kilometers from the hypocenter, and those smaller than magnitude 1.0 should be heard only in very favorable conditions, for example, during sequences of very shallow aftershocks or at the so-called swarms of shallow earthquakes where no main shock occurs.

To conclude, if you heard the sound at the beginning and during an earthquake, you were like many Zagrebians who were shaken on 20 January 2020. It was a common occurrence due to the earthquake's rather large magnitude and relative proximity.

18. See the publication by Tosi et al. (2012) in the bibliography.
19. See the publication by Sylvander and Mogos (2005) in the bibliography.

If you've heard sounds like rumblings, thunder, explosions, an approaching train, distant rumbling, rock crushing— sometimes accompanied by shaking and sometimes not—that you're sure aren't traffic or other human activity for days afterward, you haven't been alone! The sound of very shallow earthquakes is a pervasive phenomenon, not only in southern Europe but throughout the world. It would be good if these data could be collected and analyzed in more detail, which would shed more light and offer better answers to the awakened giants that many of you may have felt and heard.

5

On the crests of waves

Poseidon, the son of Cronus and Rhea, was the god of the sea, and he had a malignant nature and a short fuse. When he was angry, he was so out of his mind that he would often stick his trident into the ground in anger, causing it to shake violently. Humans and animals would then feel Poseidon's frustration like an earthquake.[1] However, according to Aristotle, earthquakes are not caused by Poseidon's anger but by unbridled winds trapped in an underground labyrinth of caves and cracks from which they try to escape.

And so, when on 28 December 2020, an earthquake with a magnitude of 5.2 struck near the town of Petrinja, 47 kilometers as the crow flies from Zagreb (fig. 3.1), few did not think that this earthquake, perhaps just like Aristotle's winds, was connected to the Zagreb earthquake from March of the same year by some mysterious and intertwined system of underground labyrinths.

Because of numerous newspaper articles after the Zagreb earthquake, even the birds on the branches of centuries-old

---

1. Perhaps unsurprisingly, the name seismology comes from the Greek words *seismos* for "shaking" and *logos* for "science."

linden and oak trees learned that this part of Europe is on the *Mediterranean-Trans-Asian seismic belt*.[2] Due to the complex interaction of the African and Eurasian plates, many microplates and other smaller tectonic units were stacked next to each other in the Mediterranean area, like parts of an ancient mosaic whose edges have been worn away by the ravages of time. Their exact shape and extent are still debated, so the plate boundary line used by USGS, shown as a red line in figure 3.3, does not continue toward Greece and Turkey. By studying numerous earthquakes in that area, we try to understand the evolution of the Adriatic microplate and its role in the seismicity of the Adriatic coast and the Dinarides, the southern branch of the Alps.

Earthquakes deeper within the Eurasian plate are quite rare. Nonetheless, there have been devastating earthquakes in history, for example, the one from 1667, which destroyed Dubrovnik and killed around 5,000 people. The Great Zagreb earthquake of 1880 destroyed many buildings in the capital, and the Ljubljana earthquake of 1895 triggered the city's urban development due to its destruction. Most of the world's seismologists will know that the Pokupsko earthquake of 1909 helped Mohorovičić discover the boundary between the crust and the mantle.[3] More recently, earthquakes in Skopje from 1963, Banja Luka in 1969, Podgorica in 1979, and Ston in 1996 are known (most of these localities are shown in figure 3.1). However, for most people in this part of Europe, these earthquakes have long been forgotten and suppressed from memory by recent war traumas.

That's why when, in March 2020, the earthquake first struck in Zagreb, and again in December 2020, in the area of the Kupa

2. See a description in chapter 7.
3. And you also know it if you have read chapter 3.

River (Pokuplje), people called upon the memories of those giants from the earlier history of this region. On the other side of the Adriatic Sea, in Italy, these earthquakes reminded the Italian seismologists and the residents of L'Aquila[4] of the tragic experience of 2009. They made many, including myself, think of the seismic waves, Mohorovičić's Discontinuity and faults again. Fault systems were often seen on the maps published in the media, one a little north of Zagreb and the other southeast, in the Pokuplje area. The fault system in the Pokuplje area is where the 1909 Pokupsko earthquake occurred, which led to the discovery of Moho (see the faults shown by black lines in figure 3.1).

Earthquakes that precede even stronger ones are rare. On average, it happens once out of 20 times within a week. Perhaps because of this, maybe also thinking that they could not be that unlucky, the residents of Petrinja and the surrounding towns in the Pokuplje area spent the next night trying not to invoke evil fate. On the other hand, some remembered the 1969 Banja Luka Earthquake that occurred on the same fault system extending into neighboring Bosnia and Herzegovina. A foreshock or twin earthquake in Banja Luka happened before the main shock. For others, the 7.1-magnitude Ridgecrest, California earthquake and its 6.4-magnitude foreshock were still fresh in memory on 28 December 2020. But, in truth, neither the seismologists nor anyone else had the slightest idea what would happen the next day.

That afternoon, at 12:20 CET, an earthquake with a magnitude of 6.3 vigorously shook the town of Petrinja and its surroundings. The one in twenty chance that nobody wanted to think about had occurred after all. In addition to human casualties and

4. For a description of what happened in L'Aquila, see chapter 7.

extensive structural damage in the Pokuplje area, the earthquake was felt in a dozen other European countries. A seismogram from the U.S. East Coast sent to me by my colleague Vernon Cormier from the University of Connecticut clearly showed surface waves. But before we get into the discussion about surface waves, let's return to the phenomenon of connected earthquakes. Namely, were the earthquakes near Zagreb and Petrinja linked, and how could they be connected given that they occurred on two different systems of faults?

The short, ad hoc answer is that they are not connected in a way that the former caused the latter. The longer answer is that they are likely not related in a way that one caused the other, but they resulted from the same field of tension or stress (or tectonic regime) in the Earth's crust in that part of the world. Due to tectonic forces, this tension field exists in the Earth's lithosphere, collectively known as the crust and uppermost part of the mantle.

The African plate is subducting under the Eurasian plate in the Mediterranean area at a speed of about 2 centimeters per year (the red line in figure 3.3 shows the plate boundary in the western Mediterranean). In addition to these two main plates, several smaller plates are in the immediate vicinity, often termed microplates. These are like a mosaic making up the Earth's lithosphere. I deliberately do not use the word puzzle because, in a puzzle, all the pieces are "locked," whereas in a mosaic, one piece can slide freely along the side of the other. Imagine placing the mosaic on a flat surface and giving it a little nudge on one side with your finger, which is the equivalent of a change in the configuration of tectonic forces. All parts of the mosaic would feel the force, some more, some less, some before, some a little later. The process is similarly applied to the Earth's lithosphere if you imagine it as a vast spherical mosaic beneath which the rocks of the Earth's mantle are buried at high temperatures. This constant interaction between tectonic plates causes tension to build up in one place and release that tension in an earthquake.

As previously mentioned, apart from reverse and normal faults, there are also faults with so-called strike-slip displacement, during which two rock blocks move horizontally, so segments on either side of the fault move horizontally next to each other. In addition to the San Andreas Fault system in California, the Anatolian Fault system in Turkey is a good example. Both of these fault systems are responsible for devastating earthquakes, like the one in February 2023 on the East Anatolian fault that killed nearly 60,000 people. From seismological and geodetic analyses, the Petrinja earthquake—in terms of how the energy was released, the way the seismic waves manifested on the seismograms, and how the surface of the Earth was deformed—was of this exact type.

It is important to emphasize that the fault systems near Medvednica and the Pokupsko epicentral area are different and separate (fig. 3.1). In other words, the Zagreb earthquake did not trigger the Petrinja earthquake. However, the problem of an earthquake forming a kind of trigger for another earthquake is very current, and many works in the scientific literature deal with it. For example, a group of colleagues from Oregon State University showed a few years ago that earthquakes of magnitude 6.5 and stronger could trigger earthquakes of magnitude 5.0 and above within three days after the first earthquake.[5] Interestingly, they published research results pointing to a marginal correlation for pairs of antipodal earthquakes—with

5. See O'Malley et al. (2018) in the bibliography.

locations on diametrically opposite sides of the Earth.[6] Somewhat more direct are studies of the passage of seismic waves from larger earthquakes, which can play an essential role as a "trigger" for smaller earthquakes in the same region. However, given that, fortunately, Croatians do not experience earthquakes as often as, for example, Californians, Japanese, or New Zealanders, they should not be particularly concerned about such a thing.

———

Because of the significant difference in magnitude between the earthquake near Petrinja and the one near Zagreb, the passing of the waves caused a significantly different sensation, which was confirmed by the first results of the seismological and geodetic data processing. As a result of excellent comments from eyewitnesses, it is now possible to write something about the destructive waves on the Earth's surface—the surface waves. But before that, let's return briefly to the body waves (the seismic waves traveling through the Earth's interior) that were felt during the Zagreb earthquake in March 2020.

With P waves, the fastest waves through the Earth's interior, the ground particles move in the direction of the wave propagation; that is why we feel them either as an impact or as a sudden pull. They have a higher frequency, act like vibrations, and are often accompanied by sound in a narrower epicentral area, especially if the earthquake is shallow. S waves are slower than P waves, they are more energetic, and the direction of ground movement is perpendicular to the direction of wave propagation. That is why, on the surface of the Earth, they are most

6. For those interested, the antipode to Zagreb is slightly to the east of New Zealand.

often felt as rocking in the horizontal plane. During the 2020 March earthquake, both P and S waves were felt near the epicenter in the broader area of Zagreb.

Štefica, from a suburb near the epicenter, said in a poll I conducted on my Facebook page that the March earthquake was preceded by a "terrible sound, a thump and a strong shudder," which can be interpreted as the almost simultaneous arrival of P and S waves at the very epicenter. Mateja, 2 kilometers from the epicenter, described the feeling of the earthquake as "sharp jolts as if someone is vigorously shaking you." A little farther away, Morana reported: "There was a rumble and immediately a strong impact, I was thrown in all directions while I was running to the children's room," and Andrea said that "the shaking was aggressive with jerks and sound." Marta. from a suburb about 13 kilometers from the epicenter, said that the building shook at a high frequency in the east-west direction, and she compared the sound to a pneumatic hammer without feeling the first impact or explosion. So, as we move away from the epicenter, the P waves gradually weaken and are overpowered by the S waves, and the sound is also gradually lost.

During the earthquake in March, one side of the fault moved upward relative to the other, so the combination of P waves and S waves polarized in the vertical direction was powerful. Kristijan wrote that the feeling of the earthquake was "pneumatic as if someone was driving the building into the ground and lifting it," and Iva described the event as follows: "The earthquake shook very quickly and violently left and right as if you were sitting on a washing machine when is a centrifuge." Petra, 8 kilometers from the epicenter, said: "First we heard a sound, a deep rumble, then strong, violent shocks, waves up and down." Mateja compared the feeling to an aggressive shaking in a basket,

Božidar to an aggressive centrifuge, and Tatjana from the sub-ridge zone to strong vertical shocks, as if "the house is bouncing up and down." Many described the movement as an aggressive rocking up-down, left-right, shaking, and strumming, while some explained the direction more specifically. All these descriptions are characteristic of a shallow earthquake on a reverse fault accompanied by an intense sensation of P and S waves. Those who contacted me from more distant places, for example, from the coastal places like Rijeka (near Volosko), Split, and the island Lastovo, and continental towns like Vinkovci (see figure 3.1), did not describe P waves, but mainly the sensation of length, which is a sign that they felt only S waves and that P waves at those distances are only recorded by sensitive seismographs.

But now we come to the main point. In the focal point of shallow earthquakes, in addition to body waves, a good part of the energy is released in the form of surface waves. Unlike body waves propagating through the Earth's interior, surface waves travel along its surface. Imagine an animation that compares seismic waves on the surface of the solid Earth with waves on the surface of the water. It consists of two cross-sections of the Earth and water, with numerous dots representing the medium through which the waves propagate. When the waves propagate through these two types of media, we can imagine what happens to the water or the solid Earth by how the dots move. Let's stare (in our thoughts) at the dot in the middle. For waves on water moving from left to right, the particle's motion is circular and prograde (oscillates in the direction of the wave motion in a clockwise direction, like a billiard ball you hit on top with a cue). In contrast, certain types of seismic surface waves are elliptical and retrograde (the particle oscillates in the direction of the wave motion like a billiard ball you spin downward, counterclockwise).

Now, let's look at the difference between the two main types of surface waves: Rayleigh waves and Love waves. As you can imagine, they are named after the scientists who first described them, Lord Rayleigh in 1885 and Augustus Love in 1911. If you could look at a slice of the Earth's volume as Rayleigh and Love waves pass through, you would see that with Rayleigh waves, the ground particles move in vertical elliptical paths, and with Love waves, they move in circular horizontal paths. When these two types of waves are combined, you can also imagine complex circular and elliptical motions in all planes. Mathematically, the early works of Rayleigh and Love describe movement at a surface in a layered medium. Indeed, the first seismological instruments installed in Europe at the beginning of the 20th century confirmed the theoretical works of Rayleigh and Love, when after large earthquakes, in addition to body waves, large amplitude surface waves appeared on seismograms.

When body waves travel from the earthquake's focus through the interior, the energy is spread out over the spherical surface, which is proportional to the square of the sphere's radius. In the case of surface waves, the energy is spread around the circumference of a circle, which is proportional to its radius. Thus, if we measure the energy of body waves first at 5 and then at 10 kilometers away from the epicenter (so at twice the distance), it will decrease four times at 10 kilometers compared with 5 kilometers, while for surface waves, it will decrease only twice. Because of this geometric property, surface waves become dominant at some distance from the epicenter. In addition to the geometric effect, body waves with higher frequencies (shorter periods) lose energy by attenuation on passage through the Earth much more efficiently than surface waves of lower frequencies (longer periods). The combined effect means that at larger distances from the epicenter, it looks as if there are almost no higher frequencies. Therefore, the sensation of P waves is often absent at greater distances, and only long-period S and surface waves survive. Gradually, the S waves weaken, and only surface waves remain. During large earthquakes, the amplitudes of the surface waves are so large that they travel around the Earth several times before all their energy is completely lost. As for the propagation speed of Love and Rayleigh waves, it is about 90% and 70% of the speed of S waves, so the following sequence of arrival of waves can be expected at a particular location: P wave, S wave, followed by surface Love and Rayleigh waves.

The depth of the Petrinja earthquake of 10 kilometers, which is the estimated depth of the focal point, that is, the beginning of the rupture, means that the earthquake was relatively shallow. It is possible that the fault's rupture continued toward the surface and that the fault line of the rupture could be seen on the surface even though confirmation from the field arrived slowly. The energy released by this earthquake was one order of magnitude higher than the Zagreb earthquake, so the surface waves that were generated were stronger. Based on the descriptions of surface wave motion, it could be expected that most comments from my survey would confirm a different, even unusual, or "strange" experience during this earthquake compared with the earthquake from March. That is why I will divide the descriptions of the Petrinja earthquake into two parts. In the first part, I will describe the diversity of motions that confirm the sensation of surface waves. In the second part, I will describe the order of arrival of various types of waves.

As for the differences in motion, many described the feeling as if they experienced waves on water. Zrinka from one of Zagreb's suburbs said: "I remember, in that whole shaking, at one moment, somewhere in the middle, it swayed as if a wave had passed under your feet, like when you are on a boat," and Ivana from another suburb gave a similar description: "Just like that, we are on a boat, left and right." Anita from a northern suburb wrote: "We were literally standing as if we were surfing on a boar." Tomislav said: "The spring earthquake was, in a nutshell, as if you were cutting waves with a ship, while this one was as if you were sailing along the waves," and Alex: "Like on a ship when the waves are rolling because of the strong *jugo*."[7] And for Mirela from Zagreb, the sensation consisted of "long-lasting waves and rocking, like in a boat at sea." Željka wrote: "The shaking went in a circle," and Renata: "It went around in a circle, it didn't swing back and forth." Lovro from Zagreb said: "Yesterday's quake lasted longer and seemed to be a mix of different frequencies in different directions." For Marijan, from a Bosnian town near the border with Croatia, the sensation was like driving an SUV on a rough forest road; for Iva, the sensation consisted of different movements, first swaying to the side and then in a circle near the end. Andreja from the center of Zagreb wrote: "The door casing danced before my eyes, the upper part towards the north, the lower towards the south," and Tomislav said: "It seemed to me that the boards were bending under my feet, like waves."

Regarding the order of arrival of the seismic waves, Štefica from the Zagreb's suburb near the epicenter wrote: "The earthquake started with a light rocking that turned into a terrible rocking of the whole house as if we were a boat rocked by strong waves." From the previous description of the motion, we can now interpret the gentle rocking from her description as S waves and the intense rocking on the water as surface waves. Zrinka from the fourth floor in Zagreb said: "There was no sound, and there was swaying, and that with quite a large

---

7. A wind that blows from the south and southeast along the Adriatic coast.

deflection." Note the absence of a description of P waves.
Mateja wrote: "In the first part of the earthquake, it was as if
there was a different shaking than in the second—I can't explain
exactly how." Milica from Split said a little further: "It took a
long time, and there was no announcement that they were coming, just rocking." Ivona from the Zagorje area, north of Zagreb,
gave one of the best descriptions of the order of arrival of the
waves: "First, a strong impact was heard . . . a light wave of rocking started, then a very short pause and then rocking again
which, according to my estimation, lasted about 30 seconds."
From the description of wave motion, here we could describe
a strong shock as P waves, a light rocking wave as S waves and
after a short pause, rocking again as surface waves. Zrinka's description also confirms this: ". . . the rocking started plus the
shaking, which intensified, and you waited for it to stop, but
it shook even harder."

Many of the comments confirmed that the feeling of an earthquake is subjective. For example, for some, even those closer to
Zagreb than to the Petrinja epicenter, the Petrinja earthquake in
December was a much more intense and eerie experience than
the Zagreb earthquake; it lasted much longer and was of greater
intensity. For others, the Zagreb earthquake was a terrifying experience not only because of the intense shaking but also because
of the eerie sound. In some locations, many described the
Petrinja earthquake as a slightly more pleasant experience than
the Zagreb earthquake, regardless of its duration. From this, the
question arises whether, apart from subjectivity, something else
affects the very sensation of the earthquake, perhaps something
that has not yet been discussed.

And indeed, many are interested in whether something else
could have influenced the different sensations in various locations. The radiation pattern diagram of surface waves and the

focal mechanism diagram (beach balls)[8] are some of the tools we have not mentioned before, which seismologists use when they have data from seismographs placed in various places and directions worldwide (fig. 5.1). From them, it can be determined that during the Petrinja earthquake, particularly expressive Love waves were released with different amplitudes in different directions. If we imagine that we look at the surface of the Earth from above, in the center of the radiation diagram circle is the earthquake's epicenter, and along its circumference are the geographical directions. The amount of energy for each direction is proportional to the distance of the point from the center. A closed line is obtained when the points in all directions are joined.

For the Petrinja earthquake, the diagram created in this way shows that the least amount of Love wave energy was released in the north, south, west, and east direction (fig. 5.1; bottom right). Most of the energy of the Love waves went in the direction of northwest, northeast, southwest, and southeast. Although the shape of the radiation diagram is common to all frequencies, the wave amplitudes are smaller at lower frequencies (longer periods). In the case of Rayleigh waves, the chart is somewhat different, and their amplitudes are slightly smaller due to the fault configuration and how the fault wings moved. But when both diagrams are made from observations on seismograms, it can be concluded that the Petrinja earthquake emitted surface waves in all directions.

---

8. Focal mechanism diagrams are also called "beach balls" because they show four areas on a sphere around the earthquake's focal point, of which the opposite two are usually marked by dark colors (compressional first motions) and the other two with light colors (tensional first motions in the seismic waves radiating outward) (fig. 5.1; top panel).

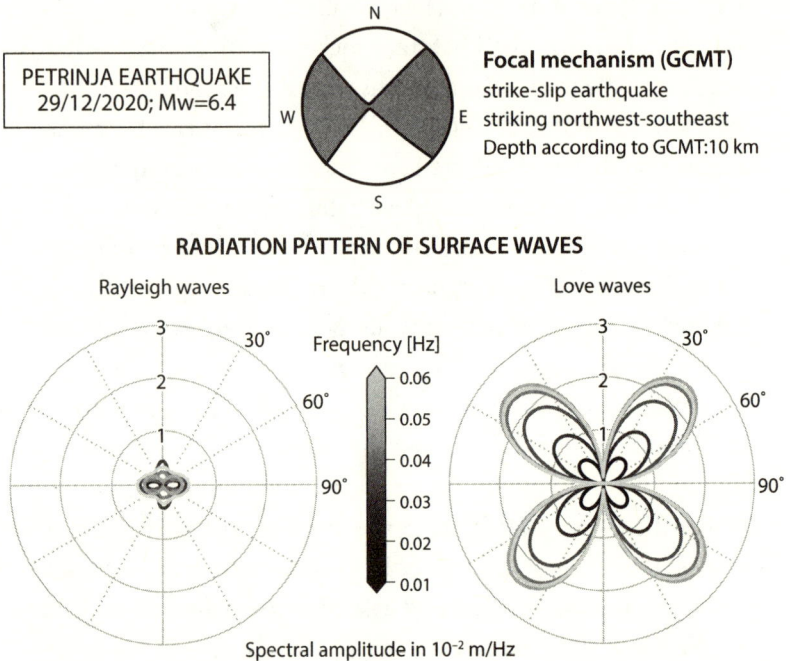

PETRINJA EARTHQUAKE
29/12/2020; Mw=6.4

Focal mechanism (GCMT)
strike-slip earthquake
striking northwest-southeast
Depth according to GCMT:10 km

RADIATION PATTERN OF SURFACE WAVES

Rayleigh waves

Love waves

Frequency [Hz]

Spectral amplitude in 10^{-2} m/Hz

FIGURE 5.1. Top: Focal mechanism or the so-called *beach ball* of the Petrinja earthquake according to the Global Centroid-Moment-Tensor earthquake catalog (https://www.globalcmt.org/). The compressional motion for various distances and azimuths from the earthquake is projected onto the southern hemisphere of the imaginary sphere around the epicenter in red, and the dilatational motion is projected in white. In this way, four separate parts of the sphere were obtained. Bottom: Surface wave energy radiation diagram for various frequencies shown in various colors.

The Zagreb earthquake happened on a reverse fault and was significantly weaker than the Petrinja earthquake. To make an adequate comparison of the radiation diagram due to a fundamentally different fault and the character of the movement of the fault wings, we can briefly mention the 1996 Ston earthquake (about 150 km southeast of Split), moment magnitude

6.0, which occurred on a reverse fault, the same as the Zagreb earthquake. In it, Rayleigh waves were emitted with a larger amplitude than Love waves. The point is that the sensation of an earthquake depends not only on the distance from the earthquake and the type of soil on which your house or building is located but also on the character of the fault and how the energy from the focal point and the fault spread on the waves' crests in various directions. The experiences of observers are the best empirical confirmation of the scientific understanding of the physical causes of earthquakes and the propagation of waves from the initiation of the earthquake through the interior of the Earth and its surface.

———

It sounds somewhat surreal that almost all the inhabitants of the broad region of central and southeast Europe, with rare exceptions, felt the Petrinja earthquake. A few days after the earthquake, the first results of geodetic methods for determining ground deformation appeared in the media. So, let's see what tools surveyors use and how contemporary geodetic methods complement seismological ones.

InSAR[9] is a geodetic method that measures the distance from a reference point, usually above the surface, to another point on the Earth's surface. In this way, the topography of the Earth's surface can be determined. However, from everything we have learned so far, the Earth is a dynamic planet, and its topography is constantly changing. Movement and permanent deformation of the ground can occur for many reasons, but in this context, we are interested in the one related to tectonic activity, manifested

9. Interferometric synthetic aperture radar.

as earthquakes and volcanoes. The passage of time, therefore, becomes a critical factor in data analysis. In simplified terms, we must measure the distance between a reference point and some other point on the Earth's surface, not once, but at different times. In our case, it is before and after an earthquake. Then, we need to look at the difference between these measurements to see how much the ground has moved in that period.

At the beginning of the method, airplanes were used,[10] where the transmitters and receivers of electromagnetic waves were located on their front and back sides. If the plane flew over a point on the Earth's surface, first with its nose and then with its tail a few milliseconds later, the distance to the target point on the Earth's surface would be calculated—using the radar method. The advantage was that the reference point for the two measurements was practically at the same place, just a few milliseconds later. Later, airplanes were replaced by satellites. The essential change in the satellite was that the time difference between the two radar images was several hours or days, the time between passes of the satellite over the same part of the Earth, instead of milliseconds as in an airplane. Since satellites have complex paths around the Earth, their distance and angle (inclination) relative to a point on the Earth's ground should be considered for each measurement. However, since satellites remain in orbit above the Earth for a long time, these measurements are continuously repeated, increasing their precision. GNSS[11] uses many satellites organized in a network. A receiver on the ground can be located with high precision when a network of satellites is employed to determine its distance. When InSAR is combined with GNSS, movements on the Earth's surface can

10. See Ulaby & Long (2014) and Uys (2016) in the bibliography.
11. Global Navigation Satellite System.

be recorded with millimeter-to-centimeter accuracy, which is exactly what we need to understand how the ground has deformed due to earthquakes.

We have already dealt with seismograms and spectrograms, so let's clarify interferograms, at least in the most basic terms. Many of you will remember diagrams from the newspapers and television in the days immediately after earthquakes, when you probably saw for the first time in your life beautiful color graphs with numerous interferometric patterns or rings, as they are called when the laws of optics are taught in school. An experienced surveyor reads what is happening on the ground from an interferogram, just as a seismologist tries to understand an earthquake from seismograms and spectrograms.

Because of the electromagnetic waves emitted by satellites, the distance is easier to represent with so-called wavelength units. For example, if the transmitter is 72 meters away from a target point and emits waves whose wavelength is 12 centimeters, we can calculate by division that the distance between the transmitter and the target point is equal to 600 wavelengths. But since the return signal must travel twice the distance (there and back), the total distance is 1,200 wavelengths. If the target point moved away from the transmitter by 4 centimeters (which would be equivalent to the ground dropping relative to the satellite's position as a result of a moderate to severe earthquake), the double distance would increase by a total of 8 centimeters, so the total distance would now be 1,200 and 2/3 wavelengths.

In jargon, it is said that there has been a "change in phase" by 2/3 of the wave cycle. However, if the target point had dropped by 6 centimeters—exactly half the wavelength of the electromagnetic waves, the total distance would have increased by 12 centimeters, corresponding to one whole wavelength and thus zero phase shift. This is called the effective wavelength because every multiple of it results in a zero-phase shift. Without prior knowledge, it would be difficult to say whether the target point deflated, moving away from the satellite, or rose, moving closer to the satellite.

An InSAR interferogram is usually displayed in wavelength units or radians between $-\Pi$ and Π.[12] There is no need to panic here because radians can be related to centimeters, as you will see below. The color spectrum shows one wave cycle from a particular reference point on a surface without deformation. This display results in rings or circles called a "wrapped" interferogram. In simplified terms, if we sum each cycle from blue to red (equal in length to the effective wavelength), we get the total displacement of the ground in centimeters relative to the ground that has not undergone deformation. A display obtained that way is called the "unwrapped" interferogram.

Wrapped and unwrapped interferograms grace the covers of scientific journals and reports for many earthquakes and volcanoes. One of them is the Cerro Azul volcano in the Galápagos Islands. The interferogram for this volcano was made based on *Envisat* satellite data for the time interval from 31 May to 5 July 2008. In other words, it shows ground deformation due to volcanic activity where the magma chamber descended. A volcano is an excellent example because its quasi-circular shape makes it easier to understand intuitively how the ground was deformed due to the emptying of the magma chamber and eventual eruption. From the quasi-circular form of the color patterns, you can see it is a volcano.

Some of the "wrapped" InSAR interferograms for the Petrinja earthquake were obtained from data from the *Sentinel-1* satellite,

---

12. Let's recall trigonometry and the unit circle. Π is obtained by dividing the circumference of a circle by its diameter, which is approximately equal to 3.14.

which is in orbit above the Earth at about 693 kilometers and transmits electromagnetic waves with a wavelength of 5.6 centimeters (fig. 5.2). It flew a little to the west of Petrinja. The previous description shows that its effective wavelength is 2.8 centimeters, which is why the displacement is expressed in the range of 0 to 2.8 centimeters. If we counted the rings on the western and eastern sides of the Pokupsko fault, there would be about ten on the western side and seven on the east side. Multiplying the number of wave cycles by 2.8 centimeters gives an estimate of a total displacement of 28 centimeters toward the satellite in the west and about 20 centimeters on the eastern side of the fault. It should also be noted that these values should subsequently be converted into horizontal and vertical motion components. When this is done, it is possible to calculate precisely how much this fault's western and eastern sides have moved.

In any case, the interferograms made for the Pokuplje area confirmed that the earthquake's mechanism was such that there was a right-lateral movement along the fault line, namely, the eastern side of the fault moved to the southeast relative to the western. In addition, the eastern side was lowered, and the western side was raised. Further, according to the slightly visible discontinuity that runs parallel to the fault line, it could be interpreted that the two sides of the fault not only moved within the Earth's interior but that the fault's rupture also probably extended to its surface. Traces could have been visible on the surface of the Earth, but different interpretations had to be taken care of because various photos circulated on the internet. Namely, not everything seen on the surface was a fault line; instead, in some places, landslides and other phenomena occurred due to the earthquake.

Let's hope that from now on, in addition to seismograms and spectrograms, you will immediately recognize wrapped and

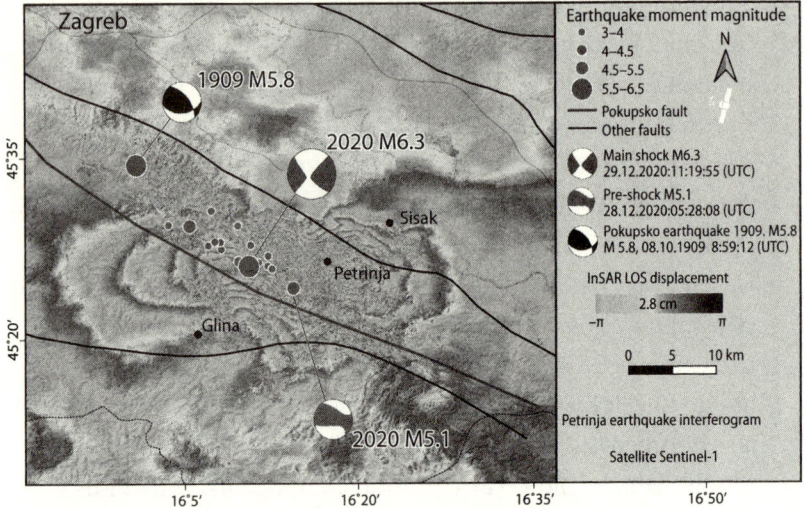

FIGURE 5.2. The wrapped interferogram of the Petrinja earthquake. The original interferogram using data from the Sentinel-1 satellite was made by Dr. Marin Govorčin from the Jet Propulsion Laboratory. The beach balls of the preshock and the main Petrinja earthquakes were taken from the United States Geological Survey, and the beach ball of the 1909 Pokupsko earthquake was taken from Herak and Herak (2010). The faults are taken from Prelogović et al. (1998).

unwrapped interferograms and beach-ball diagrams. You will search the internet for the satellite frequency band, you will know how to estimate ground displacement from interferograms, and you will not fall for conspiracy theories.

———

We have already partially answered whether the earthquakes near Zagreb and Petrinja were connected. But let us now deal with the linked (interrelated) earthquakes in more detail. The question whether the earthquakes were related to each other could be heard quite often after the Zagreb earthquake and especially after the Petrinja earthquake. Most people who have felt an earthquake know that after a major earthquake, many smaller aftershocks usually follow during the "healing" period of the fault. Of course, aftershocks, in that case, are connected to the main one. However, is there a cause-and-effect relationship between two earthquakes that occurred on two different faults?

Those interested in a more substantive answer to this question will need more reading and a brief reminder of some terms and explanations from elementary and high schools. Let us recall that an earthquake occurs when the rock mass becomes so stressed by some force that the tension (or stress) becomes greater than the internal strength of the rock, and this results in failure and sudden movement of the fault wings along the fault. From the measurement of deformation or displacement, it is possible to determine the tension indirectly. The fact is that the tension field exists in the Earth's lithosphere and is variable due to tectonic forces that continuously change over time. But what is meant when we talk about the tension or stress of rocks? To answer whether earthquakes are related, we first need to address the definition of stress and understand how rocks react to it.

As defined in physics, tension is pressure, that is, a force that acts on a unit of surface, for example, on one square meter. If you express force in Newtons, you get pressure expressed in Pascals. Every day in the weather forecast, you will hear about air pressure, because a pressure field in the atmosphere changes as a result of movement of air masses above us. The average value at sea level is 101,325 pascals or 1,013 hectopascals, called "1 atmosphere" in geophysical jargon. The less air mass you have in the column above you, for example, when you climb to a higher altitude, the less it presses on you and the lower the pressure. Thus, the air pressure at the top of the Himalayas is only about a third of the pressure at sea level. On the other hand, when you dive below the water's surface, you feel more pressure. For example, at 10 meters below the sea surface, the air pressure is approximately twice as high as at the surface. At the same time, at some point in the liquid or air, the pressure does not have a defined direction in space. The pressure in the Earth's center is more than 3.6 million atmospheres, and in the Sun's center, even around 300 billion atmospheres! The air pressure in an average car tire is slightly higher than 2 atmospheres.

This type of pressure is called hydrostatic pressure for the atmosphere and water mass and lithostatic pressure for the solid Earth. It is obtained when the weight of a column of air, water, or rock is divided by the area of the base of that column, which is also equal to the product of gravitational acceleration, density, and height of the column. We know from laboratory experiments that the strength of rocks is 10,000–30,000 atmospheres; however, in nature, when you look at a larger piece of rock, say a larger monolith, due to its inhomogeneous internal structure, irregularities, and cracks, its strength is only 100–500 atmospheres. In fact, seismologists have confirmed that a representative difference in pressure before and after an earthquake is about 250 atmospheres.

In other words, the rocks on the Earth's surface have an endurance limit beyond which they collapse under their weight. Let's take, for example, granite, whose density is about 2,750 kilograms per cubic meter. For a granite monolith to remain on the surface of the Earth, its lithostatic pressure at the base due to its weight, let's call it G, must be less than its internal strength, let's call it S. Mathematically written, $G < S$. Since $G =$ density $\times$ gravitational acceleration $\times$ height, and we know what the strength of rocks S is, from the given relation we can obtain by a simple calculation that the height of the granite block must be less than 927 meters. Indeed, the highest mono-liths on the Earth's surface are no more than 1 kilometer; for example, El Capitan in Yosemite National Park is 914 meters high from base to top.

If you understood the above paragraph correctly, you now have command of one of the tools of comparative planetology. Let's take Uranus's satellite Miranda as an example. Spectro-scopic studies have already established that Miranda is a frozen world, but because of the inclusion of other ingredients, the ice there is somewhat denser than water ice, with a density of slightly more than 900 kilograms per cubic meter. Let's take the average density of Miranda's ice as 1,200 kilograms per cubic meter. The surface gravity of Miranda is about 125 times less than Earth's, and the internal strength of such ice is about 100 atmospheres, so at least 2.5 times less than the strength of gran-ite. If we estimate the maximum height of an ice monolith on Miranda using the same formula, we get about 10 kilometers! Is it possible for such a small celestial body as Miranda to have an ice mountain higher than the Himalayas?

Fortunately, we did not have to wait too long for the confir-mation. When *Voyager 2* passed by Miranda in 1986, it photo-graphed spectacular cliffs approximately 10 kilometers high. And indeed, today, we know that the highest cliff in the Solar System is located on its very edge, on the tiny icy world of Mi-randa. It is called Vernona Rupes. I am delighted to use this example, taken from when I was a teaching assistant in the Plan-ets course at Berkeley,[13] in my current lectures of the Earth Physics course I teach to third-year physics and engineering students at the Australian National University.

13. I learned it from UC Berkeley Professor Raymond Jeanloz.

There are some critical differences between the pressure in liquids and gases and the pressure deep within the Earth. Due to the solid state of rocks, molecules do not move nearly as fast as in the air. Because of this, the stress at a point in the depth of the Earth is not distributed equally in all directions, and the lateral stress is not equal to that in the vertical direction. The African plate is subducting under the Eurasian plate in the Mediterranean area at a speed of about 2 centimeters per year, but, apart from these two plates, there are several smaller plates in their immediate vicinity, the so-called microplate, of which Adria is the most relevant for earthquakes in Croatia. This continuous interaction between the tectonic plates due to lateral forces causes the accumulation of stress in one place and a specific direction, occasional earthquakes, and the release of energy in the form of seismic waves.

Another critical difference between the atmosphere and the lithosphere is the time scale on which such changes occur. In the atmosphere, we have a pressure field in which areas of low and high pressures alternate; they stagnate or move on a time scale of just a few days. In the lithosphere, however, these changes are much slower, except when there is an earthquake and when, at least locally, the tension is suddenly redistributed. Because of all this, the question of whether two earthquakes are connected spatially is not at all trivial.

Laboratory and computer models of faults include their size, geometric shape, rock type, friction, rock pore pressure, and all the initial conditions we can imagine prevailing on a fault. It is possible to simplify things by assuming that the fault is homogeneous, that the surface is flat, that we know the value of friction everywhere along the fault, and that the initial conditions are equally given at all places.

To simplify the idea of one of the laboratory models, imagine that we have a winch connected by springs to two bricks in a row (fig. 5.3; top panel). Let us now assume that each brick represents one part of the fault, that is, part of the rock mass on one side of the fault. Of course, we can imagine more than two bricks connected by springs in a row, but two are quite enough for illustration. Let's assume that we can use springs to represent the tension in the rocks due to some tectonic force, which we can simulate by turning a winch. When the winch turns, the force tightens the spring and pulls closer one of the two bricks, which is immobile for a while due to its weight and the friction between the contact surfaces. However, when the spring becomes sufficiently taut, it is intuitively clear that the brick will move toward the winch (fig. 5.3; middle panel). This sudden movement of the brick in this thought experiment corresponds to an earthquake event in real life.

The Earth's crust is cracked, intersected by numerous faults, and consists of smaller units, much smaller than tectonic plates. When an earthquake finally occurs, the stress accumulated over the years due to tectonic forces in a particular place exceeds the endurance limit of the rocks. There is a crack, a sudden shift on the fault, and the release of energy in the form of seismic waves. When we talk about tectonic forces, we mean the forces present in the Earth's lithosphere due to the interaction between tectonic plates: due to the pressure of one tectonic plate against another because one is being pulled under the other or moving away from the other. The displacements of the tectonic plates are continuous and of the order of several centimeters per year, corresponding to the speed of fingernail growth. These constant movements cause deformations of the material, not only at the very edges of the plates but also far from the edges of the plates. The African plate slowly subducts under the Eurasian plate in the Mediterranean area. The force responsible for that is causing stress thousands of kilometers away, beneath, among others, the Adriatic Sea, Medvednica, and Pokuplje.

But what about the distribution of stress on a fault after an earthquake? Do you remember that after the Great San Francisco Earthquake and Fire of 1906, it finally became apparent that earthquakes occur on faults that already exist, not the other way around, and that earthquakes create new faults? Not so long ago, the research led to another critical realization: the stress accumulated on the fault does not "disappear" after the earthquake but is redistributed differently on the fault where the main shock occurred and in the surrounding rock mass and faults.

How the stress is redistributed after the main shock is the subject of study and is directly relevant to whether the two earthquakes are spatially related. So, let's dwell on the bricks and springs model for a moment. Let's first deal with what happens to the second brick in the row after the first is suddenly moved. Namely, now the spring that connected those two bricks is much tenser than before, and it is only a matter of time before that second brick in the row will also move. If the friction between the second brick and the substrate is not significant enough, it will move in the same direction as the first brick! From this example, we can see that when an earthquake occurred, the stress moved from one location to another, in this case, to the next position along the fault. In other words, a new earthquake will occur if that part of the fault is on the verge of breaking. If not, a sudden stress jump at the first location will not yet cause an earthquake at the second location, regardless of stress redistribution.

Of course, things are not this simple out in nature because a fault is not a line but a surface, and not a flat and smooth one but a curved surface with various irregularities. But suppose we once again simplify the complexity that exists in nature and move from a one-dimensional to a two-dimensional case. In that case, we can simulate the rock mass on a fault with a network

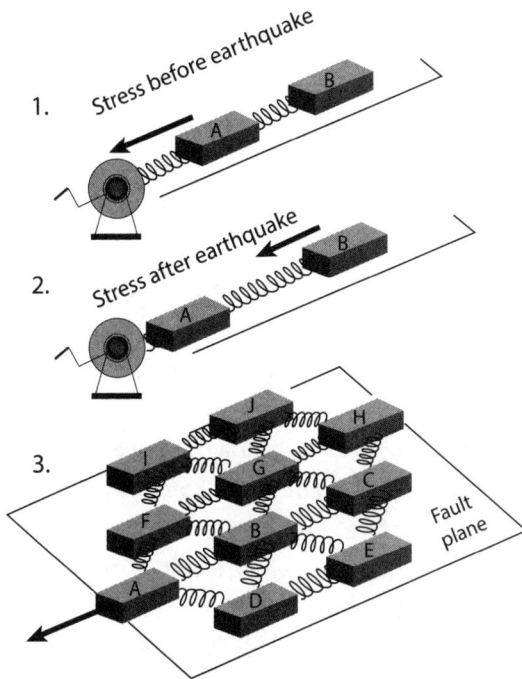

FIGURE 5.3. Schematic representation of system
tension (1) before and (2) after the earthquake
along a line. (3) Similar to 1 and 2, however, the
stress distribution is shown in the plane representing
the fault surface.

of smaller parts of the mass, say bricks, distributed in the fault
plane. Each one is connected to the neighboring bricks by
springs of a certain strength (fig. 5.3; bottom panel). The interaction between each of them is different. It depends on the initial conditions, which correspond to the density and porosity
of the rocks, pore pressure, and friction on the fault surface. If
we have prior knowledge about the characteristics of the fault,
we can simulate and evaluate how the tension was distributed

after an earthquake. That can then be used to forecast another earthquake probabilistically. Thus, after the Petrinja earthquake, forecasts made by some companies[14] based on the physical principle I have just described appeared in the media.

In addition to increased stress in some directions in space, stress will decrease in other directions. It is sometimes referred to as the "shadow of stress." Of course, one can forecast a lower frequency of aftershocks in shadow areas and a higher frequency for areas with increased stress. There are several examples of significant earthquakes in recent history that successfully forecasted where aftershocks would occur and where they would not.

Everything in the previous paragraphs refers to the spatial tension transfer after an earthquake. You may recall that the deformation on the surface, which geodetic methods can measure, and the tension in the rocks, which is not measured directly, are related. However, the laboratory model of friction on the fault also predicts that considerable deformation due to stress collection must precede the earthquake on a time scale of minutes and hours to days. There was, therefore, hope that this fact could be used for earthquake forecasting. When we touch on the "Parkfield Experiment," we will see that this is precisely why many instruments were installed to measure deformation and displacement along the fault. Unfortunately, such deformation is never clearly observed on the Earth's surface, perhaps because the nucleation zones of earthquakes are too deep, and the spatial transfer of stress immediately before the earthquake is too small to cause deformation on the Earth's surface.

Many people recall relatively recent New Zealand events when talking about the phenomenon of connected earthquakes. Namely, the earthquake in Christchurch with a magnitude of

14. For example, Temblor, Inc.

6.2 on 22 February 2011 was preceded by a strong earthquake of 7.1, the Canterbury earthquake of September 2010. Many have rightly wondered if, given examples like these, there is some pattern of behavior, some cause-and-effect relationship between the two earthquakes. Both earthquakes in New Zealand occurred on blind faults, whose existence was unknown. That is not unusual because the earthquake catalogs do not go that far back in history, given that New Zealand is a young country. Subsequent research showed separate fault systems similar to the Croatian faults: the fault system north of Zagreb and the Pokupsko fault system.

If you thought the bricks and springs model described for a single fault in figure 5.3 was complicated, you are correct. Not only is it complicated, but it is also an excellent example of a "chaotic" system, where even with essentially identical initial conditions, the results after several computer iterations begin to differ drastically. In other words, the estimation of the spatial distribution of post-earthquake stress according to the same model and for the same initial conditions is different for each calculation.

For two separate faults, for example, the fault system under Medvednica and the Pokupsko fault system, and the two faults in New Zealand, the simulations are far more complex. The current research aims to understand whether the seismic waves of an earthquake that spread through the Earth's interior can change the state of tension in the lithosphere at distances greater than the dimensions of the fault itself and even on the other side of the world. Some statistical results suggest this is so, but there is still no consensus in the scientific community on this topic.

Whether there is a cause-and-effect relationship over longer distances is ultimately not crucial, because if the rocks of one fault are on the verge of failure, an earthquake can occur on it today, tomorrow, or the day after tomorrow, regardless of whether another quake occurs elsewhere or not. Failure will happen by itself due to the continuous and relentless action of tectonic forces. Therefore, it is necessary to continue to invest in theoretical and laboratory research to achieve significant progress in understanding how the tension field in the Earth's lithosphere changes. At the same time, it is necessary to invest in the infrastructure of seismological and geodetic instruments and achieve the best possible coverage in areas with known faults.

Finally, imagine that we can represent each earthquake with a circle and then draw it on a world map. With the size of the circle, we would depict the input data on the moment magnitude. The most prominent circles would thus indicate moment magnitudes greater than 9.0. For the smallest, we would set the limit at magnitude 1.0 or 2.0. If every second represented one year, we could make a dramatic animation from the beginning of the 20th century to the present in just over two minutes. Thus, the Great San Francisco earthquake and the already mentioned Pokupsko earthquake of 1909 would appear in the first ten seconds. At the end of the first minute, the largest earthquake ever recorded would appear first, the one near Valdivia, the Chilean coast, from 1960, with a magnitude of 9.5. Not long after it, a large circle would appear in Alaska, representing the 1964 earthquake with a moment magnitude of 9.2. Among the larger circles, there would also be the Sumatran earthquake from 2004 and the Tohoku earthquake from 2011, both with a moment magnitude of 9.1. Such an animation of all recorded earthquakes in chronological order would be a perfect indicator of how dynamic the Earth is.

At the end of the animation, we would see that the earthquake locations delineate the edges of the tectonic plates, with only a few earthquakes within the boundaries they define. Although there would be periods when we would have increased activity in one part of the world, the same could not be said for the rest. We would understand that nothing exceptional is happening in the world of earthquakes today. In fact, in 2020, fortunately, there were no earthquakes with a magnitude greater than 8.0, the largest being the one in Alaska with a magnitude of 7.8. If you look at the last 10 years of seismicity in the world, you will see that 2020 was not unique in any possible way. There were nine earthquakes of magnitude between 7.0 and 7.9 (average about 15 per year), 112 earthquakes of magnitude between 6.0 and 6.9 (average about 135), and about 1,300 earthquakes of magnitude between 5.0 and 5.9 (on average, there are about 1,600). If we look at the frequency of smaller quakes and account for the earthquakes the seismographs missed, we arrive at a frightening estimate of about half a million to a million earthquakes worldwide yearly.

From all we have learned above, we can conclude that nothing special is happening to us regarding the frequency of earthquakes. Indeed, if we look at how often and with what force large earthquakes tear the Earth's upper crust, we can immediately realize that we are just a drop in the ocean, small and insignificant in the vastness of the universe. The same realization occurred when, not so long ago, in 1990, we turned the *Voyager*'s camera, which was leaving the Solar System, toward Earth and saw the image of just one pale blue dot.[15]

15. The photograph was interpreted in Sagan's 1994 book, *Pale Blue Dot*, as representing humanity's fleeting place in the universe.

6

A sharp look inside

Night descends on the southwest coast of the Barents Sea, near Russia's border with Norway, in the Murmansk Region, on the Kola Peninsula. At the height of about 100 kilometers above the sea and the frozen continent, the solar wind violently hammers its particles into the Earth's ionosphere at speeds of tens of millions of kilometers per hour. These particles then fall steeply due to the Earth's magnetic field toward the North Pole. We can follow the closed paths of the imaginary magnetic field lines in our thoughts all the way down to the core of the Earth, where the magnetic field is created and maintained.

A polar light dance begins, woven from curtains, winding lines, and rays in all shades of green. This color dominates due to the presence of oxygen in the upper parts of the atmosphere. Still, sometimes other colors of the spectrum are found in the repertoire, depending on the share of other chemical elements. With its personality, the aurora borealis fascinates the rare frost-bitten observer who gazes at the night sky. And there have been human observers here for 10,000 years since the first nomadic tribes settled on the peninsula. Who could have guessed that beneath that nightly display of light is a well-known entrance to the depths of the Earth's crust? It is the result of drilling from the time of the Cold War, when the Soviets managed to reach an incredible 12,262 meters and thus drill the deepest artificial hole in the history of humankind.

So far, the borehole on the Kola Peninsula is the deepest hole made to reach as deep as possible into the Earth's continental crust. Drilling was suspended due to much higher temperatures than expected, causing the rocks at that depth to behave like molten plastic. On the other hand, South Africa's Mponeng gold mine is the world's deepest dug hole at 3.84 kilometers below the Earth's surface. If we were to think of drilling or digging as the only ways to travel to the center of the Earth, it would, due to the lack of technology, take us from reality relatively quickly into a science fiction realm and the equivalent of a spaceship where astronauts would switch places with terranauts.[1] Fortunately, the seismic waves generated by earthquakes allow us to embark on a journey on their peaks and troughs to the deepest and most inaccessible parts of the Earth and study them just as if we had placed them under a giant magnifying glass.

In the last two chapters, we dealt with earthquakes, the giants that live beneath our feet and occasionally wake up violently, shaking the planetary interior and surface. We also touched on current research topics and learned a bit about the characteristics of seismic waves and some early discoveries that led to the "illumination" of the Earth's interior. Indeed, somewhat contrary to popular perception, when thinking about those first discoveries, we realize that not all seismologists deal with earthquakes. Many

---

1. In the American science-fiction film *The Core*, directed by Jon Amiel, a crew of terranauts heads towards the core of the Earth to use nuclear explosions to restart the rotation of the inner core, which has stopped rotating. From a scientific point of view, it is problematic that the inner core would come to a complete spinning stop and that restarting it would jump start the geodynamo.

are motivated and fascinated by the illumination of the Earth's interior. In other words, the object of their study is not necessarily the earthquake but the Earth itself, its invisible parts buried deep beneath our feet—places that cannot be reached.

However, if we know the physics of wave propagation from the earthquake focus to the seismograph placed on the Earth's surface, it is theoretically possible to penetrate the Earth's secrets. For example, we can measure how the speed of propagation and the amplitude of seismic waves change by passing through a medium, bouncing off boundaries, refracting or bending. The behavior of seismic waves is similar to light, which can tell us a lot about some distant star when we catch it with a telescope. From the analysis of the spectrum of light rays that have reached us, it is possible to determine the physical and chemical properties of a star or other distant objects. Similarly, from the characteristics of seismic waves, it is possible to examine not only the mechanism of the earthquake but also the physical properties of the medium through which the waves spread to the seismograph.

This brings us to techniques for illuminating the Earth's interior, especially seismic tomography. If you recall the discussion about direct and inverse methods when we talked about Mohorovičić's and Lehmann's discoveries, tomography could be described as an example of a modern inverse method. Moho and Inge would probably be very positively surprised by the sophistication of this approach!

Around the same time that Inge Lehmann discovered the existence of the inner core, focal plane tomography was developed by radiologist Alessandro Vallebona and refined by many of his colleagues in the years that followed. The source and receiver of X-rays had to be moved at the same time in order not to maintain exposure in one plane. The method persisted until the late 1970s, when it was replaced by much more sophisticated techniques such as computed tomography (CT) and magnetic resonance imaging (MRI). Just then, geophysicists took over the method from medical physicists and developed it further for their needs. Sources or transmitters are replaced by earthquakes, X-ray and radio waves by seismic waves, and receivers by seismometers.

Of course, in some controlled environment, you can surround a piece of the Earth's surface to hide an ore deposit from sight with a large number of explosive sources and seismographs and create an image of the underground similar to the way a radiologist obtains an image of a lung or some other organ. That is usually the case in exploration or industrial geophysics, a field with many methods and types of seismic wave sources, most often mobile. You must wonder: what if we had the entire planet Earth as a patient and, instead of active ones, relied on so-called passive seismology methods? In other words, what if only large earthquakes were available as sources that could release enough energy to be transmitted large distances by seismic waves, for example, to the other side of the planet from the source of the earthquake?

The most significant earthquakes are concentrated unevenly on our planet, mainly in the equatorial belt, where there are the most subduction zones. Farther toward the Earth's poles, earthquakes with a moment magnitude above 6.0 are rare. At the same time, seismographs are primarily distributed on the continents and much less in the world's less developed countries. So, what if we must rely only on large earthquakes and the existing seismographs?

To illustrate the problem a global seismologist faces due to the uneven distribution of seismographs on the Earth's surface and the locations of large earthquakes, let's take the example of a medical physicist who needs to calculate radiation doses for brain CT. When everything works as intended, radiation with a thin beam of X-rays from multiple directions is achieved by rotating the source around the patient's head. The sources of X-rays are located in the circular part of the CT machine, which you enter in a lying position as though entering a giant white doughnut. When X-rays pass through tissue, their energy weakens depending on the density of the tissue. This attenuation can be measured and represented by the so-called attenuation coefficient, and its value tells you about the tissue density. Research has shown that the attenuation coefficient changes depending on the tissue type, for example, for various types of cancer. When the rays pass through the tissue, the CT machine uses digital sensors positioned opposite the source instead of film. From them, the data are sent to a computer, which creates a series of digital images of cross-sections along the brain. The word tomography comes from the Greek word *tomos*, which means precisely "cross-section." The radiologist then interprets these cross-sections and makes a diagnosis from the variations of the attenuation coefficient.

But let's imagine now that the medical physicist and radiologist do not have access to a correct CT device. Let's envisage, for example, that the rotation of the source is limited and that, for certain angles, the sources simply do not work. In other words, let's say that the very thing that makes a CT scan better than a simple X-ray—the ability to irradiate the brain from multiple angles—doesn't work as intended. In this case, the medical physicist could adapt to the new conditions and calculate the radiation doses to consider the problem's new geometry. However, computer algorithms in their original form would not be suitable for creating tomograms. Such a situation would resemble the situation we have in global seismology. Namely, due to the uneven distribution of large earthquakes and seismographs on the

Earth's surface, we are forced to seek compromise solutions and be innovative to image the interior of the planet Earth.

We can draw at least one more parallel between tomography in medicine and in geophysics. Seismic waves passing through the solid Earth weaken, just as X-waves do on their way through the human body. This weakening of the wave strength due to attenuation is clearly visible on the seismograms as a weakening in the amplitudes of the recorded time series. It is intuitively clear that the characteristics of the material through which the seismic waves pass can then be determined from these data. For example, from the Earth's attenuation, we can infer something about its density and temperature and, a little more indirectly, its structure.

In addition to attenuation, seismologists also measure the arrival time of waves and, thus, the speed of propagation of waves through the medium through which they pass. To get a better idea of wave travel times, let's take the example of longitudinal (P) waves, which spread like sound from an earthquake source on one side of the Earth, through the core, to its opposite side. These waves take about 20 minutes to travel from the source through the very center of the Earth and reach its other side. From the model of the Earth that we have based on observations of hundreds of thousands of longitudinal waves, the precision with which we can predict the travel time is of the order of one tenth of a second. Thus, the mean value of the travel time of longitudinal waves from a given point on the Earth's surface to its antipodal point on the other side of the planet through the very center is 1,212.1 seconds.

However, some waves move faster than average, and some move slower. Now, let's imagine that we color the paths of waves through the Earth that are faster in blue and those that are slower than average in red. For every major earthquake, we

have thousands of seismograms at our disposal, as well as a catalog with the exact focal time of the quake. We can measure the
times of occurrence of P waves and, thus, their travel times
through the Earth. There are already published catalogs with
measured travel times of P waves for each earthquake and each
seismological station. If we took the travel times of thousands of
waves and transferred their deviations from the average to the
Earth's interior in shades of blue and red, we would end up with a
blue-red conglomerate of the Earth's interior. We could then make
numerous horizontal or vertical cross-sections, just as if you
were cutting a *Gugelhupf*[2] with a knife from the surface towards
the center. These sections are called tomograms. One such tomogram showing slow (red) and fast (blue) areas of the lowermost
mantle at its boundary with the core, is shown in figure 6.1.

In reality, tomography is not so simple because we cannot
assign the same color to one wave path through the entire volume of the Earth. This is because, along that path, the waves will
encounter slower or faster material, similar to the car you use
to travel from one city to another; the quality of the road will
determine its speed. To use the algorithm to determine the
speed of the waves in each part of the volume that we want to
study, it is necessary to target it with as many waves as possible
from different directions.

How this unknown volume is expressed, that is, divided in
the tomographic problem, represents a technical challenge. Depending on the type of problem, it is usually divided into cubes
or spherical prisms, and the volumetric coverage of waves from
all directions dictates their size. When the coverage is better,
like near the Earth's surface, where we have more recorded
waves from local and regional earthquakes, we can divide the

2. A cake baked in a ring pan, popular in the central parts of Europe.

FIGURE 6.1. A tomographic map of the lowermost mantle projected to the core-mantle boundary. Red areas are slower, and blue areas faster for the PKP waves than average. Ray paths of PKIKP waves from large earthquakes (locations shown by red spheres) through the Earth's interior to seismological stations (green spheres) in a three-dimensional view focusing on the Pacific. Note extensive seismogenic zones surrounding Australia, from the Indonesian earthquakes in the north to the prominent Vanuatu-Fiji-Tonga-Kermadec (in clockwise direction) earthquakes in the northeast. The rays of seismic waves connecting the earthquakes with the seismic stations are shown in color, where red corresponds to slower and blue to faster-than-average values. The Earth's mantle is transparent for illustration purposes. The snapshot is from an animation created by Vizlab National Computational Infrastructure in collaboration with the author. The tomography was presented in the publication Tkalčić et al. (2002).

volume into many smaller parts and thus achieve a higher resolution. On the other hand, the problem we face with the deeper parts of the Earth is the lack of their coverage by wave paths.

This volume division in tomography is called model parameterization. Recently, we achieved an advance in that the model's parameterization is not explicitly given but is determined as an unknown. That is done by dividing the space in a very special way using the so-called *Voronoi cells*.[3] Translated into layman's language, it means that we do not assume that we know anything about how and into how many parts the volume of the Earth we are exploring should be divided, but a computer algorithm does it for us based on the data. If the problem is well-posed, the algorithm will assign a seismic wave speed value to each part of the volume, a shade of blue for faster and red for slower waves.

What can we learn from Earth tomograms? First, the Earth's interior is not as homogeneous as the interior of an avocado. In a sense, it is more like a pomegranate. From tomograms of the Earth, we can, to put it simply, learn how geological structures are arranged in space, similarly to how we can image tumor cells in the human body. Although tomograms are only snapshots of the Earth's interior at this moment in time, they allow us to understand the Earth's past and present, that is, how it evolved from the accretion of the planets until today. For example, the

3. In the simplified case, we can have randomly distributed points in the plane, or "seeds" in mathematical jargon, and around each of them construct polygons so that every point belonging to a polygon is closer to the seed of that polygon than to any other seed. The polygon lines are thus always strictly between some two seeds. Polygons defined in this way are called Voronoi cells, according to mathematician Georgy Voronoi. Interestingly, many shapes in nature look like Voronoi cells, for example, the shapes of the structures on the fur of a giraffe, the lines on the wings of dragonflies, tree leaves, garlic heads, and bee combs. If you look at crystalline structures or epithelial cells under a microscope, you will notice, you guessed it—Voronoi cells!

Earth's crust is highly heterogeneous, and these heterogeneities change on many spatial scales. From smaller sedimentary basins through mountain ranges and their bases, all the way to ocean basins and continents. Seismic tomography for the upper mantle clearly shows subduction plates with fast wave speeds, as blue fingers, drawn into the ambient weak red. Some plates stagnate at the boundary between the upper and lower mantle at depths of about 660 kilometers, while others penetrate the lower mantle down to its lowermost boundary with the liquid core.

The lower mantle of the Earth can be imagined as a graveyard of lithospheric plates and the place of birth of hot material that expands and ascends like a giant mushroom. Two huge, red volumes with slow wave speeds stand out in particular—one under Africa and the other under the central Pacific (see figure 6.1). Although the convection in the mantle is slow, and it takes hundreds of millions of years for a fraction of the volume from the lower mantle to rise to the Earth's crust, the tomograms somewhat resemble synoptic maps of the pressure or temperature fields.[4] The red and blue zones offer information about the location of geological structures in a similar way to the interpretation of the results of CT, magnetic resonance, or X-ray images by a specialist doctor.

While we can certainly say that today, we are at the stage of mapping the planet's surface, we still know very little about its deep interior. The main reason for this is that there is relatively much less data than for the upper parts of the Earth's interior. Near the surface, we can use numerous smaller earthquakes and even microseismic noise due to the interaction of the ocean,

4. For example, see Mousavi et al. (2021) for tomography of the lower mantle from S waves and Muir et al. (2022) for tomography of the lower mantle from P waves; both are listed in the bibliography.

atmosphere, and solid earth. However, to illuminate the deep interior, we need large earthquakes, large enough that waves from their focal points penetrate to the bottom of the mantle and core and, after that, return to the surface. In addition to being weakened due to attenuation, these waves must also pass through the upper layers of the Earth on their way. Only when we know the structure of those upper layers, can we properly correct wave travel times to explore the deep interior. For all these reasons, the tomographic image of the Earth's deep interior is still relatively hazy.

However, we are slowly improving that picture with more sophisticated seismographs, more powerful computers, and better inverse and numerical methods. In fact, to solve some problems, we have to step away from tomography because it still does not have the resolution to distinguish all the structures in the Earth's mantle. Of course, apart from tomography, seismologists have developed numerous other methods. One of them is not an inverse but a direct approach, and the recipe for it is pretty simple:

(a) The seismogram describing ground motion at point A due to the earthquake from point B is recorded.

(b) A theoretical (synthetic) seismogram is calculated based on an understanding or prior knowledge of the structure between points A and B and the propagation of waves between them. Remember, we called it the "model" of the Earth earlier.

(c) Compare the observed and synthetic seismograms. Imagine that one of the two seismograms you have is printed on transparent paper so you can compare the seismograms by overlapping them. If the ground oscillations shown by the squiggly lines overlap, in mathematical terms, we say that the adjustment or "fit" is good. Of course, the overlay is never perfect, but a computer algorithm can calculate and express in numbers how good or bad it is. If we are satisfied with the fit of the synthetic to the observed seismogram, we accept it together with the Earth model we used to calculate the synthetic seismograms.

(d) If we are not satisfied, we change the Earth model via its parameters in the next iteration. For example, if the theoretical seismogram is slower than the observed one, we need to tune the Earth model to increase its wave propagation speeds. We recalculate synthetic seismograms based on the new Earth model. We then compare the theoretical and observed seismograms and keep doing so until the fit is good enough to accept the Earth model as the final model.

This method is called "forward modeling" of the waveforms recorded as the seismogram. We use it as a deductive method to study the regions of the Earth's deep interior through which seismic waves have passed, for example, after waves have passed through the lower mantle or bounced off the core-mantle boundary. Using this method, we obtained by modeling the characteristics of the deepest layers of the mantle direct evidence that the oceanic lithosphere ended up at the boundary between the liquid core and the mantle on a path several thousand kilometers long and lasting several hundred million years. We concluded that on that boundary's upper surface, apart from the previously mentioned two huge domains of slower material under Africa and the Pacific, there are heterogeneous structures with a size of about 100 kilometers or less in diameter.[5] These are dimensions invisible to tomography, whose resolution in the lower mantle is only about 1,000 kilometers.

5. See Li et al. (2021) in the bibliography.

For some of these heterogeneities, we showed that they got there through material recycling. In contrast, others are probably the remains of the primordial material that has remained there since the planet's formation. But how can we know that? We got the answer when we jointly analyzed seismograms with geodynamic simulations.

So, let us stay about halfway to the Earth's center, at a depth of about 2,900 kilometers, where the silicate mantle, whose internal material moves very slowly, meets the liquid core of iron and nickel[6] with more vigorous internal movement of material. Precisely at these depths, in some places at the boundary between the mantle and the core, seismic waves suddenly slow down. That is quite unusual, because the increased density in the Earth's mantle normally increases the speed of seismic waves continuously with depth. Global seismologists in the last century noticed these structures and called them ULVZs.[7] Since the boundary between the core and the mantle is the most extreme discontinuity within the Earth, more extreme even than the Earth's surface,[8] it is perhaps not entirely unexpected that these most enigmatic planetary structures are located there. However, what is extraordinary is that in a relatively thin layer of only a few tens of kilometers, which sits on the upper side of the core, the speed of the waves drops to half the value, and the density increases by a third!

At first, it was thought that these mysterious structures were connected with hot spot volcanism at the Earth's surface, among which Iceland is one of the more famous examples. It

---

6. See McDonough and Sun (1995) in the bibliography.

7. *Ultra-low velocity zones.* Ultra-low here means very low. See, for example, Williams and Garnero (1996) in the bibliography.

8. In terms of the density contrast.

was thought that the structures might be a deep source for this type of volcanism, anchoring partially melted mantle at the bottom of the mantle. However, it turned out over time that there is no spatial connection between the ULVZ and hot spot volcanism. We recently went a step further and concluded that these regions of the lowermost mantle are most likely made of material that differs in its chemical composition from the surrounding mantle.[9] We were able to do this thanks to advances in digital seismological data analysis using sophisticated *Bayesian transdimensional inversion* methods.[10] Namely, with these innovative methods, we could peer into the interior of the ULVZ, which is impossible with computer tomography because we do not yet have the resolution for that. We have shown that these zones consist of layers. In simple terms, we achieved increased resolution using inverse engineering, that is, a waveform modeling method by which you set up a model of the Earth's structure, simulate waveforms based on that model, and then compare them with observations to improve the mathematical model of the Earth until the predictions and observations agree. That requires modern algorithms of mathematical geophysics and millions of hours on many supercomputer processors. And

9. See Pachhai et al. (2022) in the bibliography.

10. Statistical methods that use Bayes' theorem from 1763 to calculate the probability of an event or hypothesis based on available data or information and update it each time new data or information becomes available. For example, we can calculate the probability of a later event, given that the previous event has occurred. Transdimensional here refers to changing the number of so-called free parameters in the inversion, i.e., unknown characteristics of the model that you are trying to determine by modeling. In other words, the total number of free parameters with which a model is parameterized or described in the inversion is not fixed but becomes a free parameter in itself. If you would like to know more, see the publication Sambridge et al. (2013) in the bibliography.

what does this discovery tell us about the layering of the mysterious ULVZ structures?

To answer this question, we need to recall the first chapter and the accretion of the planets. More than 4 billion years ago, when the heavier chemical elements sank toward the center of the Earth, and the lighter ones differentiated in the Earth's mantle and crust due to buoyancy, one of the larger planetesimals, Theia, struck the proto-Earth. The Moon was formed from the molten material that stretched due to the collision. Its internal structure and rotation still bear the hallmarks of that event and serve as the strongest argument in favor of this hypothesis. Due to the increase in temperature and melting of materials in the interior of the proto-Earth, the heavier chemical elements sank toward its center.

We can imagine a great ocean of magma at the bottom of the mantle. It consisted of solid and molten minerals, rocks, gases, and crystals trapped in magma. Heavier elements that sank to the bottom of this ocean gradually formed layers, just as is the case with the formation of sedimentary rocks near the planet's surface. During the following billions of years, when the mantle began slow convection, the layers that formed at the bottom of the mantle were pushed due to this convection into smaller piles by lateral movement in the lower part of its cells. These piles are higher than the highest mountain ranges on the surface of the Earth, and we have come to know them as zones of extremely low seismic wave speeds.

Through geodynamic modeling, we showed that it is possible that the chemical heterogeneities brought to the bottom of the mantle by the dynamics of planet formation very early in Earth's history remained there as buried time capsules that did not mix with the surrounding mantle. The results are compatible with the existence of the layers that we observed by analyzing the seismograms. So, by the heterogeneities that we see with the help of seismology, we mean morphological and chemical characteristics that change in space. In addition, at these depths, there are such high pressures that the properties of atoms are different from those on the surface. They are packed as much as possible into crystal structures that look quite different for the same chemical composition closer to the Earth's surface.

There is a hypothesis that the oceanic crust, which is subducting into the Earth's mantle, is melting, which may also explain the origin of the ULVZ materials. Namely, we saw that, according to the theory of plate tectonics, the oceanic crust is recycled over time through the mantle by subduction to the bottom of the mantle and the rise of hot magma to the surface on the mid-ocean ridges at the edges of the tectonic plates. There are also places in the middle of the plates where magma rises through the mantle in tubular structures and finds its way to the surface, pours out as volcanic lava in hot spots, for example, Iceland or Hawaii. However, as we have already mentioned, there is no spatial correlation between ULVZs in the lower mantle and such volcanoes on the Earth's surface. This supports the hypothesis that the mysterious ULVZ structures are, after all, accumulations of primordial material.

To finish with the innermost shell of the Earth—its core, here the picture is also getting sharper, perhaps not yet with tomography due to the poor coverage of that part of the Earth by body waves, but certainly with numerous studies of travel times and waveform modeling, focused on those parts of the core for which we have the data. With this method, we recently discovered that longitudinal or P waves move faster through the inner core in the direction of the Earth's rotation axis than in the equatorial plane.[11] This is evidence of anisotropy, that is, the

11. See Poupinet et al. (1983) in the bibliography.

property of a material to behave differently for seismic waves in various directions.

We also found another ball inside the inner core with a different anisotropy than the rest of the inner core.[12] We call it the innermost inner core. Since anisotropy can tell us how iron crystals are oriented and how they were oriented when they crystallized, this fossilized crystallographic structure, most likely due to the sudden change in the direction of crystal growth, is the subject of numerous studies. This structure could be due to a different crystallographic phase of iron stabilized at the innermost inner core conditions.[13] It could also be associated with a change in the growth regime of the inner core, the re-establishment of geodynamo and the magnetic field, the birth of plate tectonics, and even the trigger for the sudden development of life on the Earth's surface.[14]

More and more data and their analysis brought us the first images of the boundary between the inner and outer core—it seems that it is more heterogeneous than we initially thought, which could be due to different crystallization rates from the liquid to the solid core in various places. In turn, that could result from the complexity of convection in the liquid core and how it affects heat flow from the inner core to the outside. For example, suppose the molten iron sinks in a column of the outer core cooler than the surrounding material. In that case, the flow of heat from the inner core to the outside at the bottom of that column will be increased due to the larger contrast in

12. See Ishii and Dziewoński (2002) in the bibliography.

13. There is an ongoing debate on the stable phase of iron in the inner core. See, for example, Belonoshko et al. (2017) in the bibliography.

14. The innermost inner core could be the fossilized remnant of a long-ago global cataclysm, the Cambrian explosion, but also the result of gradual physicochemical processes in the core that we do not yet fully understand.

temperature at the boundary of the two cores. That will create conditions for faster crystallization of the inner core at the spherical front of its growth so that the material's morphology will differ from the surrounding one. Seismic waves will "feel" all these differences and bring information to the surface.

From all of the above strands of information, it can be concluded that the inner core is truly a complex world whose secrets we manage to penetrate in part thanks to modern global seismology.[15] It is a planet within the planet, its internal engine and magnetic heartbeat; it is what gives life but also takes it away. We will return to this topic in the last chapter when discussing the core and the magnetic field of Mars, which died off at one point in Martian history.

The farther we look into the Earth's interior, the sharper the images we can obtain. The challenges of structural studies attract to the field of geophysics a large number of computer experts, physicists, and mathematicians who want to solve practical problems. Most of all, some are interested in how to sharpen the image we have in an optimal way, that is, how to achieve the best possible results in as little time and with as few computing resources as possible. That is important because tomography is an iterative process in which calculations need to be repeated many times. The muscles of supercomputers around the world are being flexed to achieve higher resolution. A higher resolution is indeed and remains one of the primary goals of global seismology. However, not the main goal, because it would be like saying that the ultimate goal of astronomy is to sharpen the images of distant objects in space.

---

15. For more details, see the books *The Earth's Inner Core: Revealed by Observational Seismology* (Tkalčić 2017) and *Earth's Core* (Cormier et al. 2022) in the bibliography.

7

Dragon's jaw
and crystal ball

The city of Tangshan, with more than a million inhabitants, was sleeping soundly on a warm summer night. The unbearable heat of those days at the end of July 1976 was unprecedented. Twins Da Feng and Fang Deng had spent a good part of the previous afternoon in their modest apartment on the second floor, wiping the sweat from their foreheads right next to the fan their caring father had managed to procure for them earlier that day. They giggled, touching the protective netting with the tips of their nearly identical seven-year-old nose tips, as they stared, mesmerized, up close at the edges of the propeller blades merging into a closed circle. Watching them in that carefree game, Yuan Ni thought about how much she wanted another child.

A little farther from this family idyll, along the railway line that passed through the center of Tangshan, a truly unusual scene unfolded that Tuesday. A dense swarm of dragonflies[1] followed, like a dark cloud, the last carriage of the train slowly

---

1. There are more than 6,000 dragonfly species in the world, of which 1,000 are found in China.

moving from the direction of Luanxian to Tianjin. People who got off their bikes on the ramp to miss the train let them fall instinctively onto the road and shielded their faces with their hands.

It was another busy and sultry day in industrial Tangshan, and many fell asleep late with wide-open windows. At exactly 3:42 a.m., a mighty hidden fault beneath the city awoke and revealed its existence in the most monstrous way possible. The earth thundered, the sky sparkled, and the horizon turned purple. Underground and surface waves instantly shook and bent the railroad tracks like toys, and water pipes burst under the force of the horizontal movement of the ground. Brick houses and buildings fell one after the other like towers of cards. A scream broke through the night. In just 16 seconds, Tangshan turned into a continuous ruin as far as the eye could see. The narrow passages between buildings and streets were soon flooded with naked human bodies covered in dust and blood, and moans and cries became a reality.

Most brick buildings without supporting elements collapsed after the first few seconds. The building where Yuan Ni's family slept was luckily only slightly better built than the surrounding ones, so it resisted the violent horizontal movement for more than ten seconds. But at one point, a heavy concrete beam broke away from the structure and nailed the bed where Da Feng and Fang Deng were sleeping firmly to the floor. A gap opened beneath them, and the bed collapsed along with the walls and contents of the room they were sleeping in.

When Fang Deng regained consciousness, she heard surreal sounds of sobbing and crying around her. She couldn't utter a word from the shock, and she couldn't move from the weight of the burden above her. Daylight was coming through the crack, and she could make out the sound of her mother's crying that spread through the outside air and mingled with the men's voices and the sound of the helicopter in the distance. One of the voices spoke to the mother: "We are sorry, a large beam pressed your children, the son on one side, and the daughter on the other. We can barely move the beam on one side and save only one of them. You will have to choose who to save."

Once again, Fang Deng heard her mother crying and begging the unknown men to save both of them. Her tiny body was utterly broken and powerless. She tried unsuccessfully to open her mouth and call her brother on the other side, whose arm was crushed by the beam.

Outside, on a pile of bricks and planks, Yuan Ni continued to plead with this unplanned, hastily assembled team of rescuers to save both of her children. A little farther down the road, her husband's motionless body lay among the dead. She looked up at the sky as if searching for an answer. Above her, numerous parachutes with packages of food and first aid slowly descended from the height toward the epicenter of this tragedy. They were a confirmation of the harsh reality, if anyone was still hoping they were only dreaming. And then, under the time pressure, because every second was precious to the rescuers who were already calling to another place, completely devastated, she uttered in a broken but audible voice, "Save my son . . . save Da Feng." These words hit little Fang Deng's heart deep in the ruins of the building like a heavy arrow, worse than the earthquake itself and the knowledge that the end was near.

That is how the Chinese film *Aftershock*, directed by Feng Xiaogang, begins.[2] As a complete antithesis to the special effects of Hollywood, this story, from the ruins of Tangshan, takes

2. The film was a historical success as the first IMAX film made outside the United States.

us straight to the depths of the soul to that part of it that explores suffering, survival, anxiety, death, love, and forgiveness. From the moment of this Sophie's choice,[3] if it can be called that, this saga is an emotional roller coaster that describes human destinies determined by the earthquake and keeps you riveted to the screen until the very end.[4] It's an understatement to say that those 16 seconds of terror and what followed changed someone's world. *Aftershock* described it faithfully, powerfully, wonderfully, and layered, without expensive special effects and exaggeration. If you've been lucky enough to avoid an earthquake until now, the film will most likely change everything you thought you knew about the human tragedy caused by earthquakes.

Will earthquakes be predictable?—is the question of all earthquakes-related questions. Because let us be realistic and completely honest: no matter how interesting the research on the Earth's interior is, whether it is about plate tectonics or the magnetic field, convection in the mantle or the inner core, this is a question that is on everyone's lips, it all starts and ends with it. A question that cries out for an answer.

When we say prediction, we mean three elements: the exact location, time, and magnitude of the earthquake. We haven't succeeded in this yet, but will we be able to in the future? Numerous

3. Taken from the title of a film by Alan J. Pakula from 1982 based on William Styron's 1979 novel of the same name. To save at least one of her two children from certain death in the concentration camp at Auschwitz, Sofia sacrifices her younger daughter.

4. Those who have seen the films of another director from China, Zhang Yimou, may not be surprised. If you decide to watch *Aftershock*, it could be etched in your memory as the best movie about earthquakes as natural and human disasters you've seen so far. I would highly recommend that you prepare tissues, even if you are one of those people who never cry at a movie.

books have been written about forecasting, scientific and political debates are held, and courts are in session. Scientists break a lance over it. To understand this problem—its very core—let's take a step back and stay a little longer in Asia.

————

You can often read rude and even vulgar comments under on-line newspaper articles about the purpose of the seismological profession when people in their post-earthquake trauma realize that seismologists don't forecast like meteorologists do, such as forecast hail or tornadoes with high accuracy. An approximate answer to these comments could be given with the following targeted question: "We still can't beat malignant diseases, but should we stop researching because of that?" Let's now look back at the topic of predictions from the perspective of some historical facts and see which direction today's research is going.

We are used to discussions about earthquake causes after every event, particularly in the places where the world's earthquakes occur. There are discussions about their frequency, and quite often, there are those who claim they could recognize the coming earthquake in something else. Whether it's a full moon, a planetary conjunction, too much rainfall, the neighbor's cat, bone pain, overexploitation of the planet's resources, or greed, people tend to believe that earthquakes have simpler explanations than physical forces in the interior of the Earth and, of course, that they can be predicted.

Those who have learned something about plate tectonics cannot escape thinking about how much and how they are bound to this planetary process by their position on the globe. This applies to one of the countries lined up on the Mediterranean trans-Asiatic seismic belt, from the Mediterranean to Turkey, Syria, and Iran, or in Japan, Indonesia, New Zealand, Chile, California, or

Alaska on the "Ring of Fire" around the Pacific. Thus, sifting through the gray matter of our hippocampus, searching in our mind for countries around the Ring of Fire, we can easily skip over the most populous country in the world. China, with a large part of its territory, has slipped inside the Eurasian plate, and indeed, it does not immediately come to mind as one of the countries near the edges of the plates, most affected by earthquakes. A good part of it consists of stable blocks of continents—cratons—which have defied subduction and tectonic forces for billions of years. The same forces pull the plates under each other and drag them deep down into the bowels of the Earth, all the way to the boundary of its mantle with the core.

However, it is a historical fact that the far east of Asia, notably the North China craton in the east of China, which was formed about 2 billion years ago, was torn apart by monstrous earthquakes for centuries—to the extent that six of the ten deadliest earthquakes in recorded history happened right there. The worst was the Shaanxi earthquake of 1556, from the time of the Ming Dynasty, which claimed about 830,000 lives. Many faults in this part of the world are ancient and inactive. Still, they have been reactivated from time to time, due primarily to the forces transmitted to northern China through the continental mass from the Himalayas, where the Indian plate presses against the Eurasian plate. Add in the massive subduction along the Japan-Kurile oceanic trench near the eastern edge of the Eurasian plate, and you will immediately understand why the interior of the Earth is restless in this part of the world, even when it is thousands of kilometers away from the edges of the tectonic plates.

That is why it is not without reason that China was where the first seismoscopes, primitive devices for recording ground movements, were invented. The first seismoscope, from the time of the Han Dynasty, was the invention of Zhang Heng, consisting of a bronze urn on the outside of which were attached dragon heads facing the eight cardinal directions of the world. Frogs with wide open mouths were arranged under each head. Every time the ground shook due to an earthquake, the dragons would drop pellets from their throats that would end up in the mouths of the frogs on those sides of the world where the motion was most pronounced. The design of the urn's interior is unknown, but it is speculated that the mechanism that activated the balls that came out of the dragon's mouths was a kind of pendulum located in the middle of the urn. It would react to ground motion caused by earthquakes.

Some 18 centuries after Heng's invention, during the Cultural Revolution, when many city youths were sent to be reeducated by peasants, much historical literature, sculptures, paintings, and architecture were destroyed. The question is whether Heng's original seismoscope, even if it had been preserved in mainland China among the rare important relics that did not end up in the national museum in Taipei after the civil war, would have survived the cultural heritage demise that followed in the last phase of the revolution. However, historical writings and replicas have survived, some of which were made at the beginning of the 20th century.

Be that as it may, after about three centuries of relative seismic calm, northeast China was rocked by several strong earthquakes in a time interval of only three years: in 1966, the Xingtai earthquake of magnitude 7.2; in 1967, the Hejian earthquake of magnitude 6.7; and in 1969, the Bohai Sea earthquake of magnitude 7.4, in the extreme interior of the bay within the craton. Scientists have noticed that the hypocenters of these earthquakes migrated northward over time. They considered that they are a sign of the activation of a complex tectonic system, where the Eurasian plate begins to deform significantly due to the configuration of forces in compression to which it is exposed.

The belief that smaller events precede strong earthquakes was quite widespread in China, which is essential for our prediction story and gives a good picture of the mindset of the population and also of the Chinese seismologists of that time. One of the reports after the annual scientific convention in Beijing in 1972 speaks of smaller earthquakes preceding earthquakes of magnitude larger than 6.0 by a year or two in advance. Therefore, it is not surprising that earthquake education intensified in China after the triple event sequence, and the authorities soon began teaching the local population how to record observations of groundwater levels, underground electrical currents, animal behavior, and anything else that could be used to predict earthquakes.

In June 1974, the first of the two official predictions of long-term earthquakes in Haicheng was published. The publications were followed by increased seismic activity, a drop in underground water level, the emergence of snakes from their winter dens, geodetic anomalies, and other phenomena. Then, short-term forecasting led to the evacuation of the population, and finally, an earthquake of magnitude 7.6 occurred in the evening hours of 4 February 1975. Unfortunately, this was the first and last time that an earthquake has been predicted successfully, at least according to several criteria and the judgment of the broader scientific community. But let's step back into the past a bit first.

In those days, numerous commissions sat in China, and the document produced in the middle of 1974, after the already mentioned Beijing Conference, was particularly important. Several locations were explicitly named where earthquakes of magnitude 5.0–6.0 were expected in the next year and a half. However, a more extreme opinion was also represented, that of the northeastern part of China having the potential for an earthquake of magnitude 7.0–8.0. This opinion refers to climate data and comparisons and similarities of the situation at that time with the period before significant historical earthquakes. Although cautious about prediction, the report was an important psychological factor in China at the time because the State Council supported it. From there, it was transmitted in a pyramidal manner down to the levels of provinces and autonomous regions, then to the management of cities and prefectures, municipalities, communes, and, finally, to production brigades and teams, that is, individual families.

Apart from the major earthquakes of the late 1960s, geodetic leveling prompted long-term and medium-term forecasting, which measured the height difference between two fixed points several hundred meters apart. Those measurements revealed an anomaly; namely, the annual height difference changed by only 0.1 mm, but then its value changed by as much as 2.5 mm in only ten days. In truth, no one could exactly explain what such a sudden steep change between two points could mean. Theoretically, it could have been caused by increased pressure and compression of lithospheric blocks, perhaps even bending the lithospheric plate away from its edges. Another thing crucial to the medium-term prediction and the psychosis that followed in late 1974 and early 1975 was the frequency of minor earthquakes, which increased considerably. On the other hand, some further observations, which at first seemed strange, to say the least, for example, variations in magnetic field intensity and deviations from average tide values, were later found not to be anomalous compared with long-term averages.

———

With a population of 1 million at the time, Haicheng was only a tiny municipality in Liaoning province, in Chinese terms. It was the middle of winter, the coldest month of the year, with an average temperature of −10 degrees Celsius. This province was part of the occupied territory called Manchuria during the Second World War and the scene of one of the biggest battles of the civil war after the liberation. Liaoning province became the industrial center of China after 1948, leading in steel production. Unfortunately, due to the presence of numerous coal and iron mines, it soon received the epithet of one of the most polluted.

The air in this part of the world is dry as gunpowder in January, and somewhere away from the steel mills and mines, above the golden wheat fields, the star-strewn night sky must look spectacular. But when the January winter creeps into your bones in Haicheng, even *baijiu*—a white spirit with a base of sorghum and sometimes rice, wheat, and barley, with at least 35% and often up to 60% alcohol—does not help either.[5]

When Yuan dressed warmly early in the January morning and headed from his brigade along the frozen road to Dashiqiao on the shores of the Bohai Sea, he could not have known that he would find an unusual sight. At first, he thought it was from drinking too much baijiu the night before. But as he approached, it became apparent that two completely frozen snakes of the genus *Lycodon*, or king snakes, as they are popularly called here, lay motionless by the side of the road. As it was later established, it was just one of the unusual observations of these snakes, which must have been jolted out of their winter sleep by something and driven out of their warm underground den into the freezing cold outside.

---

5. By the way, with more than 10 billion liters sold per year, this alcoholic drink is the most sold and, at the same time, the most consumed alcoholic drink in the world. Its volume exceeds the combined whiskey, vodka, gin, rum, and tequila markets.

Although since the beginning of 1975 the frequency of minor earthquakes had decreased significantly, the number of so-called macroscopic observations—everything from rising groundwater levels to the strange behavior of various animal species—had grown rapidly. For example, the well-water level dropped by more than half a meter; according to eyewitnesses, the water became cloudy and changed its taste. In addition to more than a hundred independent sightings of snakes, sightings of frogs and rats, as well as other anomalies in animal behavior, were reported. Although I don't have the details of the report, I suspect that according to some, catfish also became agitated, hens stopped laying eggs, and geese became more belligerent. Whatever you may think, animal behavior as an omen of earthquakes was nothing new or unique to the Chinese. According to written documents, it originates from ancient Greece, from the 4th century B.C., when it was recorded that snakes, weasels, and centipedes left their dens days before devastating earthquakes.

But even if we disregard other observations of strange animal behavior and probably some questionable reports, the emergence of numerous snakes from their dens in the middle of such a winter is a truly unusual phenomenon. Amid the psychosis that reigned in the people, it was something that could only strengthen the belief that it was an animal instinct to survive a catastrophic earthquake rolling somewhere under the hills.

It was not known precisely what mechanism or perhaps simply an extended spectrum of vibration senses was involved. But on the evening of 3 February 1975, a series of smaller earthquakes began, growing to several hundred by morning. The sequence culminated with an earthquake of local magnitude 5.1 on 4 February, shortly before 8 a.m. Zhu Fengming, head of the Liaoning Seismological Service, was confident that a relatively large earthquake was in the offing when he submitted a situation report to Hua Wen, a high-ranking military figure in the Liaoning Revolutionary Committee. Although no one in the Service knew precisely how to account for the numerous macroscopic observations or believed that earthquakes could be predicted in the short term, after a series of smaller night and morning tremors, they instinctively sensed that something much bigger was brewing.

After a conversation with the Seismological Service, probably alarmed by the night and morning earthquakes, Hua Wen made a critical decision on 4 February, at 8 a.m., to sound the alarm and evacuate the population. As it turned out later, this decision would save thousands of inhabitants from certain deaths, and in later analyses, it would be shown as a successful model of short-term earthquake prediction. Finally, at 7:36 p.m. local time, a magnitude 7.3 earthquake on the Jinzhou fault began violently shaking Haicheng and the entire Liaoning province. Less damage was also recorded in Seoul, Korea, and it could have been felt in the Primorsky Krai in Russia and Kyushu, Japan. Although it later turned out that the evacuation did not go smoothly and perfectly, about 2,000 dead and 27,000 wounded are still relatively low figures compared with the 150,000 estimated number of dead if nothing had been done.

After the Haicheng earthquake, there was a growing enthusiasm and belief that Chinese forecasting methods could be applied more widely. Some have reinforced the belief that big earthquakes come in winter or early spring. Others were inclined to believe that the systematic monitoring of seismicity and other geophysical parameters and the behavior of animals weeks and months before an earthquake would play a key role, according to which every earthquake would be destined to be predicted in time, and people would be adequately prepared for it.

———

But then, on 28 July of the following year, in the middle of summer, and without prior notice in the form of minor earthquakes or previously recorded strange animal behavior and other anomalies, the crowded neighboring province of Hebei was hit by the Great Tangshan earthquake, a tragedy that would leave the world speechless. While the little girl from the beginning of this chapter miraculously survived, the consequences of the quake would be catastrophic.

On that day, Tangshan, entirely unprepared for the earthquake, relived the Shaanxi tragedy of four centuries earlier. According to official data, the earthquake that shocked Tangshan and the whole world that night had a moment magnitude of 7.6. The very center of Tangshan, an oblong zone along the southwest-northeast fault, about 10 kilometers long and about 5 kilometers wide, roughly along the railway line, was subjected to the unimaginable intensity of shaking at level XI. On the scale of intensity, the only case worse than XI is XII, a natural disaster in which there is a massive change in relief. Tangshan shook violently because of the fault lying just beneath the city, the thick sediments, and the type of construction that did not allow for sizeable earthquakes. More than 242,000 people lost their lives, with more than 164,000 seriously wounded. However, figures three times higher have also been mentioned. In 7,218 households, all family members died. More than 160,000 families lost their homes, and more than 4,000 children became orphans in less than half a minute.

Tangshan was razed to the ground, and all access roads were destroyed; it was left without water and electricity and was cut off from the rest of China for a time. Traces of liquefaction and sand fountains were visible everywhere. Without rescue teams ready to respond immediately after the earthquake, after numerous aftershocks—the largest of which was a magnitude 7.1 on the fault northeast of Tangshan—thousands of those who remained alive under the rubble after the main shock succumbed to their wounds. Even in Beijing, 110 kilometers away, 50 people died due to the intensity of the shaking. One hundred thousand soldiers, 30,000 construction workers, and 30,000 medical personnel participated in the rescue and reconstruction, and Chinese officials refused international aid.

To go back a little further, the story that spread after the Haicheng earthquake took on a political dimension of its own, although it went in an opposite direction from that of L'Aquila, Italy, where seismologists were accused of misleading the public about the possible danger of an earthquake.[6] While in Italy, seismologists were sued, here, they were raised on a pedestal. In the historical context in which the Haicheng earthquake took place, when the leadership of the Liaoning Provincial Party Committee was glorified, and when the news reached the rest of the world in a trickle, such a political connotation should not be surprising. Some details were either missing or overemphasized. The rulers in China were portrayed, surreally, as people who held the keys to the fate of the proletariat in their hands. As those who, among other things, could predict earthquakes.

---

6. The 2009 L'Aquila earthquake killed 309 people and devastated the city. Although in 2012, six scientists and one government official were initially sentenced to six years in prison for manslaughter for, according to the prosecution, spreading inaccurate, incomplete, and contradictory information about the possible danger before the earthquake, an acquittal was handed down in 2014. The Supreme Court confirmed it in 2015. While the American Geophysical Union and many other world organizations defended the scientists, others believed that the verdict was not pronounced against science and the impossibility of predicting earthquakes but against the failure to properly communicate scientific information to the public.

Realistically speaking, there was a medium-term forecast of the Haicheng earthquake, albeit for a much smaller earthquake than the one that occurred. Still, it is difficult to argue that there was also an official short-term forecast of a few hours. Apart from the sudden increase in seismicity from the previous day, there was no established scientific method by which one could conclude that an earthquake would occur. Nonetheless, the importance of amateur macroscopic observations cannot be denied in this story. Likewise, no one can deny that thousands of people were saved by timely evacuation.

But after the Great Tangshan earthquake, enthusiasm for earthquake prediction died abruptly. It was as if someone in a loud audio equipment store had turned off the main switch, and all the sound mixing suddenly died down. Once again, in the history of seismology—as after the Great San Francisco earthquake—a turning point occurred, and this relatively young science took a new direction. Instead of relying on earthquake omens, forecasts, and predictions, it set out to explore the Earth's internal architecture and understand the physics of earthquakes. Instead of decisions based on macroscopic observations and monitoring of seismicity, emphasis was shifted to improving the understanding of the Earth's internal structure, the distribution and state of faults, the propagation of seismic waves from the focus of the earthquake to the surface, and on improving the mitigation of the effects of earthquakes on people and buildings. However, let's see if this approach has brought us any closer to answering the question of whether earthquakes will one day be predictable.

———

Perhaps the story of these two relatively recent earthquakes, from Haicheng and Tangshan, has prompted you to think about human motivation and the pursuit of solutions against natural disasters, especially earthquakes. This desire for their predictions is probably as old as the first successfully lit fire of some wretched but bright *Homo erectus*, who sought salvation from the winter in the caves of Africa, in *loess*[7] hollowed out by the desert wind. Many who have survived the horror of the earthquake need no additional motivation to understand this. However, it was necessary to redirect the methods and mindset of scientists from the time before and after the Tangshan earthquake and all those who thought forecasting would be easy. It is no coincidence that this story took place in China. If reading about the Tangshan earthquake or watching the movie *Aftershock* caused you to shed a few tears thinking the outcome was terrible, then perhaps you should remember the most terrible earthquake in recorded history, the 1556 Huanxian (or Shaanxi) earthquake, with a moment magnitude of 8.0.

The Shaanxi earthquake occurred in northern China, near the border with Mongolia. As mentioned, according to various estimates, 830,000 people died in it! Imagine how many people that is for the 16th century, for a time when the population of the whole world is estimated at slightly less than 500 million. Practically 1 in 500 inhabitants of the world disappeared in a single earthquake! It was an earthquake in the interior of the Eurasian plate on the normal fault, which separates the Weihe ridge from the Ordos block. In some places, the scars of this earthquake can still be found on the slope of the Ordos block, up to 8 meters wide.

Most people in that part of China at that time lived in dwellings hollowed out of loess. Loess is abundant there due to the Gobi Desert and the action of the desert wind that brings and deposits fine sand crystals from the north. In Chinese, this type

---

7. See the glossary of relevant terms for the explanation.

of dwelling is called a *yaodong*. Life in the yaodong was in more or less the same conditions in which *Homo erectus* lived a million years ago, perhaps improved by some bronze vessels, a clay ceiling, and stone-clad exterior walls. More terrible than the magnitude of this earthquake is the fact that due to the force of the shaking, the yaodongs immediately collapsed, closed, and thus became tombs for hundreds of thousands of people.

In addition to destroying countless homes, the overnight Tangshan earthquake also shattered dreams about the ability to predict earthquakes. It showed that it is necessary to accept that earthquakes, caused by the internal dynamics of the planet, are extremely nonlinear phenomena, that each one is different, and that there are no reliable forecasting methods in the sense that you can predict their exact time of occurrence, location, or magnitude. While in one case, minor earthquakes precede the main one, there is no announcement in the other. Or maybe there is, but at that time, no one was clear what the announcement was, or what kind of announcement could describe common characteristic of all earthquakes. Seismology had to take new and different paths. There was work to be done. This brings us back to California in the 1970s and 80s, to a small, picturesque town of only 18 inhabitants—Parkfield—located between San Francisco and Los Angeles, near the central part of the San Andreas Fault.

You're probably wondering why Parkfield. In addition to a small bluegrass festival and dancing farmers, this small town in the shade of large oaks is known to the seismological world for its turbulent geological history. Namely, on average, significant earthquakes have occurred in Parkfield every 22 years since the middle of the 19th century. But it was fascinating that the recorded seismograms for the earthquakes of 1922, 1934, and 1966 were almost identical, one wiggly seismogram line to the other.

In addition, the 1934 and 1966 earthquakes had foreshocks—about 17 minutes before the main shock—whose seismograms also looked very similar.

You wonder how such a thing is even possible. Such similarity of seismograms is possible only if the same fault surface is always activated and recorded with the same instrument at sufficiently long waves.[8] Of course, the shorter the waves, the greater the differences. In other words, you have a source—an earthquake and a receiver—a seismometer at fixed locations and waves propagating between them through the same material. So, you have a perfect natural laboratory and an experiment set up in it. You just have to wait long enough. In this case, it was part of the fault between the northern section of the San Andreas—where the Pacific and North American plates slowly "creep" or "crawl" past each other, little by little, a few millimeters a year—and the southern part of the San Andreas, which is practically locked, without the slow movement of the fault wings along the fault.

Scientists, therefore, had good maps in hand to investigate the mechanisms of earthquakes that recur from time to time on an active, well-monitored fault. Since the mid-1980s, they have installed a whole arsenal of instruments near Parkfield and along the fault: powerful seismographs, then strainmeters, which measure rock deformation at a depth of 200 meters along the fault, magnetometers for measuring the intensity of the magnetic field, creepmeters, which measure displacements on the surface along the fault, and other scientific "weaponry." They forecasted with 90%–95% confidence that the next earthquake there would occur between 1985 and 1993. Some of the key questions were:

8. See also a discussion about repetitive marsquakes in chapter 10.

(a) How is stress distributed in space and time on the fault due to the action of tectonic forces before and after the earthquake?

(b) Do earthquakes repeat at an average time interval, or is each earthquake unique, a story in itself?

(c) How do the structure of faults and surrounding rocks affect the nucleation of smaller earthquakes and the possibility of larger ones and their distribution in time and space?

They wondered what the deformation we measure on the surface could tell us about the stress distribution on the fault, and they hoped for a positive result—confirmation of the predictions for earthquake occurrences between 1985 and 1993.

They waited and waited. I found myself among that team, who in those years, apart from the Lawrence Livermore National Laboratory, worked once a week with colleagues at the USGS California office in Menlo Park, in the northwestern part of Silicon Valley.[9] Indeed, the USGS and Berkeley were the epicenters of expectations because the scientists there were the most involved in the story. Although I did not work on topics related to the "Parkfield Experiment," I could hear first-hand what colleagues involved in it were doing and thinking. A lot has passed since 1993, and it was clear on the faces of the USGS seismologists, and you could feel it in the air that many of them had somehow overcome that disappointment.

Eventually, a magnitude 6.0 earthquake did happen in Parkfield, but not until 2004. We greeted the most watched and studied earthquake in human history with a huge question

9. At USGS, I worked with my colleagues Bruce Julian, Gillian Foulger, and Doug Dreger on the Icelandic earthquake with unusual radiation of seismic energy.

mark above our heads; it occurred 11 years after its forecasted time. Devastating. That's why the "Parkfield Experiment" left a bitter taste of disappointment in the mouth. But, as they say, only those who dare to fail eventually succeed. Research continued.

Why is earthquake prediction so tricky? Each fault is different—some of them we know about, but many we don't—earthquake catalogs don't go back far enough, and, after all, underground architecture is entirely invisible to us. We do not know how deep the fault reaches, whether it is a flat or curved surface, whether its surface is smooth or rough, whether and where it touches other faults, the chemical composition of the rocks on one and the other side of the fault, or their physical properties, for example, strength and porosity. We do not know precisely how the deformation we observe on the surface of the Earth can be related to the deformation and stress in the depth of the fault. We also do not know many other factors. A forecast can be made, but by its very nature, it must be probabilistic and taken with a grain of salt. So, how do we proceed?

Not everything is so negative. The first good news is that seismic hazard maps exist in most countries. They are well made, but of course, they must be constantly updated. The other good news is that, based on fundamental knowledge of physics and the propagation of seismic waves through the interior and across the surface of the Earth, we can predict how the ground and some buildings will behave during an earthquake, and that is already a major benefit. This is possible because of basic science and seismological research on the nature of the subsurface, in a similar way that radiologists can illuminate the inside of the human body. Ironically, earthquakes help us because they serve as a source of waves illuminating the Earth's interior. It is possible to predict infrastructure behavior during earthquakes due to the development of engineering, construction, computer science, and numerical methods. Either way, those hazard maps serve as input for engineers, builders, and insurance companies.

In the end, the most positive thing is that modern studies involving laboratory models and artificial intelligence are being carried out across the world, aimed in the direction that one day we will be able to predict earthquakes. Certainly not without major investment in science and technology, which will need to continue to develop. This might even take us to the point where we will have to place thousands or millions of microsensors on every fault in the Earth's interior and then monitor the strain in real-time. In a way, we will have a "crystal ball," an insight into the dynamics and future behavior of faults. In fact, we are already doing it today, but we have only scratched the surface of the Earth with the help of satellites. InSAR, LIDAR[10] and GPS[11] are just some of the networks and methods that give us an insight into where the Earth's crust is most stressed from surface deformations.

The stress or tension build-up mechanism on a fault is still under investigation. It is most likely that the hot rocks of the Earth's continental crust beneath approximately 15 kilometers of depth are ductile, and this rock mass "flows" at a higher speed than on the surface, but without earthquakes, and the upper part of the crust therefore bends and the stress along the fault surface increases. However, how this stress is distributed in space is not yet known.

Furthermore, laboratory experiments at high pressures and temperatures give us insight into how hard rocks are and how

10. Light detection and ranging.
11. Global positioning system.

strain and stress are related. The chemical and physical structure of the soil is examined by drilling around the fault. Old tree trunks are explored, and excavations are made to detect historical earthquakes on rock samples.[12] Investments are made in studying the deeper interior of the Earth and the mechanism of earthquakes using seismic waves and tomography methods. Investment are also made in mathematical geophysics, as well as in machine learning and improved techniques for processing enormous amounts of digital data. Investments are also made in alarm systems based on the detection of P waves. Even a few seconds of warning before the arrival of S waves can be crucial to saving people and infrastructure. Likewise, investments are being made in modern construction resistant to earthquakes.

The conclusion is that we need to learn to live with earthquakes, just as probably a billion people on the planet, including those who live in some of the most attractive parts and the most beautiful cities of the world. These lie along the edges of the continents and ocean coasts near the edges of tectonic plates, volcanoes, and earthquakes. Unless you want to move to stable parts of the continents, somewhere in Siberia, to the northernmost, permanently frozen parts of Canada, or the remote regions of the Australian Outback seldom struck by earthquakes, you must learn to live with them.

12. That is what *paleoseismology* deals with.

8

Red

Ascending the Stuart Highway from the direction of Alice Springs toward the north, I leave the cross ridges of the Macdonnell Ranges behind me. The winding road gradually turns into a straight line leading to infinity. Slowly at first, but as the signs become less frequent, I move north faster and faster. The Sun is high above the horizon. Tires stick to hot asphalt. Vehicles coming from the opposite direction become less frequent until encounters with them turn so rare that, according to the Outback's unwritten rule, occasional drivers greet each other by gently lifting one hand from the steering wheel and giving a significant wave. Cockatoos, ravens, eagles, and the first termite mounds soon appear, slowly turning into tall castles toward the north. I'm leaving the Tropic of Capricorn[1] on the southern side. Behind the wheel, the feeling is similar to when you are sailing out to sea behind the wooden helm of a boat. My destination is the Warramunga[2] facility about 500 kilometers north.

There is something magical, not easily explained, in the redness of the Australian Outback. It seems surreal, and it permeates

1. See the explanation in the glossary of relevant terms.
2. Land of the Warumungu or Warramunga Aboriginal group near Tennant Creek.

your whole being. It's all around you, from your iris and pupil to wherever you can see through the windshield and farther beyond the horizon. You know it's there too, thousands of kilometers away. Even the darkness seems different when you know that redness surrounds you. No one is near you, sometimes for hours, especially when you set off before dawn. When you are alone with yourself like that, when you realize that, to some satellite, you are just a lonely bright dot slowly crawling in a straight line in the heart of the continent, you have a lot of time to think about your life's path and calling. What main roads, side ways, and invisible detours brought you to this road?

In my case, what would have happened if one morning, while I was still a postdoctoral fellow, one of the world's few preeminent seismologists, Brian Kennett, had not come to my presentation in Santiago de Chile, suggested that I visit Canberra and consider applying for a faculty position to continue my academic career at the Australian National University? What if the interview and selection hadn't ended as they did? What if I hadn't registered for the conference in Chile? What if I hadn't even worked on the projects I came to present? Instead of a PhD in seismology at Berkeley, I could have pursued another professor's offer for a PhD in geodynamical modeling of the convection in the Earth's mantle. I could have chased other opportunities and made different decisions. If only one had developed differently, would I have ever ended up on this road? Who knows, maybe as a retiree on a tourist tour, such as the route from Darwin to Adelaide by the train Ghan,[3] one of the offers that can still be found in the repertoire of world travel agencies.

3. Its full name is Afghan Express, after the camel drivers from Afghanistan who were among the first explorers of the Australian continent in the 19th century.

The proliferation of seismographs and seismological equipment around the world had not been a topic close to me since my first days in science, when, like other PhD students in the narrower field of solid Earth geophysics, I spent hours and hours in front of a computer monitor. We came from various fields of science, from physics and applied mathematics to astronomy, engineering, computer science, and geophysics. We all had in common a background in physics, mathematics, and informatics and an inclination to data analysis, computers, programming, and numerical modeling. If any candidate without that background were to rush in here, their learning curve and catch-up must have been extremely steep.

I enjoyed coming to the Berkeley Seismological Lab in the morning, sitting down in front of my favorite Sun Microsystems computer, and staying engrossed until the wee hours in my FORTRAN and Matlab programming routines, Unix scripts, and a series of commands written in the *Seismic Analysis Code* and *Generic Mapping Tools* applications,[4] which were spinning numbers from day to day. We downloaded the digital data of ground motion records from the internet or got them to use from local seismograph networks. We then analyzed these waveforms using the existing tools or developing our own.[5]

Fieldwork was somehow foreign and distant to me. In my mind, it had always been done by someone else. Admittedly, as a person trained in radiotelegraphy during my mandatory military service, I was no stranger to sitting in an SUV supplied with communications equipment on some slope of the Julian

4. See the references Goldstein and Snoke (2005) and Wessel et al. (2013).

5. Among the others, I shared the lab with Charles Mégnin, Michael Pasyanos, Yuancheng Gung, Akiko Toh, and Junkee Rhie, friends and seismological colleagues with whom I am still in close contact although we live on four different continents.

Alps and sleeping in a tent with a stormy sky descending. But I imagined technicians from the Seismological Service, and casual, well-paid workers went to this kind of fieldwork. At the same time, science lived in a computer laboratory, in moments of inspiration and labyrinths of microprocessor integrated circuits. The only field-going persons I genuinely envied were Neil Armstrong and Buzz Aldrin, the two astronauts who installed the first seismograph in the lunar Sea of Tranquility during their "fieldwork" in the summer of 1969. I would definitely have signed up for that type of fieldwork!

At that time, when I was a PhD student, we had about a hundred global broadband seismographs that reliably recorded the seismic wavefield wavelengths from about 100 seconds (0.01 hertz) on one side to about 0.02 seconds (50 hertz) on the other side of the spectrum and were available almost in real-time. Each seismograph was just a dot on the world map, from historical locations such as the San Francisco Bay area—the site of the first installed seismograph in the Earth's Western Hemisphere[6] to distant and exotic oceanic islands, African savannas, and Antarctic ice.

What interested me then were the paths of waves from large earthquakes through the Earth's interior to the existing seismological stations scattered around the globe. Many of those stations were established as part of astronomical observatories because, from the early days, it was in the interest of astronomers to correct telescope observations for the oscillations of the solid Earth after great earthquakes. Indeed, I downloaded

6. The first seismograph installed in the Western Hemisphere was at the Lick Astronomical Observatory on Mt. Hamilton in 1897, and its director, E. S. Holden, from the University of California at Berkeley, believed that all earthquakes should be recorded to achieve better control of the position of astronomical instruments.

the data I used in my doctorate on the Earth's inner core from the internet with the help of several centers that collected and distributed data to the world scientific community.[7]

Besides the ancient Chinese seismoscope story,[8] I remember several anecdotes about seismographs. I heard one related to Andrija Mohorovičić, a founder of the Seismological Station in Zagreb in 1906, from Marijan Herak while I was still a student. On the newly installed seismograph in the basement of the Meteorological Observatory on Grič (fig. 3.4), more precisely, in paper number 9, Mohorovičić recorded the Great San Francisco earthquake![9] I heard another anecdote from Bob Uhrhammer, a legendary seismologist from the Berkeley Seismo Lab, from whom I learned much in my first scientific steps. It was about a sheriff in the Mexican state of Puebla who shot a Benioff-type seismograph with his revolver to stop one of the earthquakes there!

Finally, I have fond memories of visiting the historic College Outpost seismological station, which has been run by the University of Alaska in Fairbanks for many years and is momentous in seismology. I experienced my *Northern Exposure*[10] and the aurora borealis to collect local data from the Alaska Seismological Network, which was unavailable online then. They meant a lot to me because of the specific paths of seismic waves that "traveled" here from the South Atlantic. Namely, these waves

7. In particular, the Incorporated Research Institutions for Seismology (IRIS) and their Data Management Center (DMC) for archiving and distributing digital seismological data stand out. I am grateful for their enormous effort and professionalism. See the link to IRIS DMC in the bibliography.

8. See the previous chapter, *Dragon's jaw and crystal ball*.

9. See chapter 4.

10. Popular television series (1990–1995), *Northern Exposure*, is about life in a small town in Alaska.

spread from the focal points of large earthquakes near the South Sandwich Islands,[11] through the inner core of the Earth, all the way to the far north, where seismographs in Alaska record them.

My personal experiences with fieldwork and installing seismographs were thus limited to a few visits to the recording stations during my doctoral and postdoctoral days in California. Now and then, I needed information that wasn't available online, and that required me to learn a bit about what was happening behind the scenes. On one occasion, I managed to get my hands on records of nuclear explosion tests, which were collecting dust on outdated magnetic tapes. Their digitization led me to observe the waves from a single event that passed twice through the Earth's center by ricocheting from the Earth's surface on the opposite side from the earthquake.[12] The waves were recorded on a seismograph near Mina in western Nevada, in the vicinity of nuclear tests. It took them about 40 minutes to return to the same location after the blasts were triggered. Exotic waves after large earthquakes and explosions have been recorded before, but never for two passes (there and back) through the very center of the Earth, which was my project's goal. Thus, amazingly, seismic waves originating from blasts on

11. Near these islands in 1915, the ship *Endurance*, an expedition led by Sir Ernest Shackleton with the aim of crossing Antarctica on foot, was trapped in the ice. The ship was found in early 2022 at the bottom of the Weddell Sea.

12. For the first time, I managed to isolate the waves that spread from the nuclear explosion through the very center of the Earth to the antipodal point on the other side of the globe, where they bounce off the surface and return through the interior and center of the Earth to the epicenter itself, where a seismograph records them. With colleagues Megan Flanagan and Vernon Cormier, we were able to show how these waves bounce off intriguing structures in the Earth's upper mantle. See the publication Tkalčić et al. (2006) and the book *The Earth's Inner Core: Revealed by Observational Seismology* (Tkalčić 2017) in the bibliography.

one side of the Earth led to new knowledge about the Earth's inner core and the properties of the Earth's lithosphere on the opposite side of the world from where they bounced back.

In conclusion, I could say that, apart from simple gratitude for all the hard work that took place somewhere behind the scenes of scientific discoveries and pursuits, in that first part of my scientific journey, I had no experience setting up seismographs in the field.[13] But if you want to gain a good understanding of what's hiding deep beneath a point on the spinning globe that you've stopped with your index finger, you need data. You need recorded ground motions, which we have already learned much about in previous chapters.

Some instruments are permanently and expertly installed at the bottom of small underground vaults shaped as concrete pipes, wide enough only to climb down with a ladder. Typically, the civil service takes care of them. These real-time data are used to monitor earthquakes and tsunamis; however, there aren't many such instruments on a continent like Australia, where you'd need a helicopter to maintain most of them. Many more seismographs are required for better coverage and academic research, so you must install them yourself in the field. Therefore, by coming to Australia, I got a rare opportunity through my scientific work to visit some of the most remote parts of the planet, where humans have hardly set foot.

Indeed, Australia is sometimes referred to in geophysics as an astonishing natural laboratory because it is surrounded by the largest seismogenic zones in the world. From Indonesia and Papua New Guinea in the north, through the Philippines, Fiji,

---

13. However, things took a somewhat unexpected turn when I accepted an offer from the Australian National University and a professorship in the Research School of Earth Sciences in the Seismology and Mathematical Geophysics group.

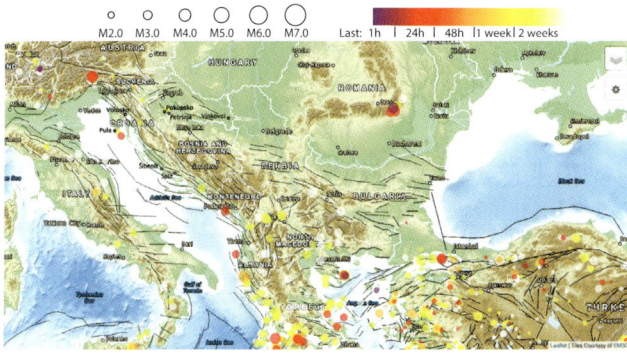

FIGURE 3.1. Two weeks of seismicity in the Mediterranean in January 2025, with most earthquakes on the Apennine and Balkan peninsula (notably, in Greece) and Anatolia (Turkey). Only magnitudes 2.0 and above are shown, with the size of the circles proportional to the magnitude. Major faults are displayed with black lines. The most prominent fault is the North Anatolian Fault in Turkey, running roughly east-west. Note many smaller faults and their orientation crisscrossing the area. The localities of the cities and towns in Croatia and the surrounding countries mentioned in the text are also shown on the map. The colors show the chronology of the earthquakes. This map is produced using the catalog and mapping tools from the Euro-Mediterranean Seismological Centre at https://www.emsc-csem.org/.

FIGURE 3.3. Ten years of seismicity (2012-2022) centered on the Apennine and Balkan peninsula. Only magnitudes 2.5 and above are shown, with the size of the circles proportional to the magnitude (see the legend). Note that most larger earthquakes are overlayed by the subsequent smaller aftershocks, so they cannot be deciphered. Major tectonic boundaries are displayed with red lines. This map is produced using the catalog and mapping tools from the United States Geological Survey (USGS) at https://earthquake.usgs.gov/earthquakes/map/.

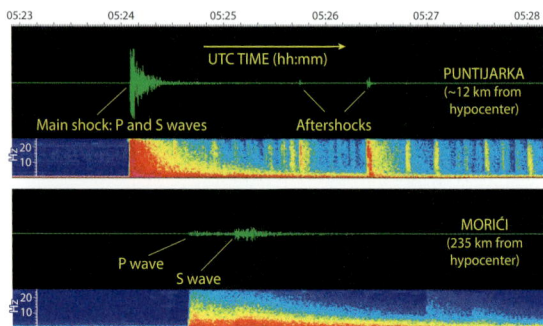

FIGURE 4.1. Seismograms and spectrograms for the 2020 Zagreb earthquake, recorded at the seismological stations 'Puntijarka' (top two diagrams) and 'Morići' (bottom two diagrams). The horizontal axis of the seismogram shows time, and the vertical axis shows the speed of moving ground particles during the passage of seismic waves. The horizontal axis of the spectrogram shows time, and the vertical axis shows the frequency of ground particle oscillations. The colors show the strength of the frequencies - stronger frequencies are shown in red, and weaker frequencies are shown in blue. At the Puntijarka station near Zagreb (Figure 3.1), P and S waves arrive almost simultaneously, and at the Morići station near Šibenk (Figure 3.1), the arrivals of P and S waves are visibly separated. The original seismograms and spectrograms were graphically presented by Dr. Marija Mustać Brčić from the Seismological Service using the 'Scream' software (see the page: https://www.guralp.com/sw/scream.shtml.)

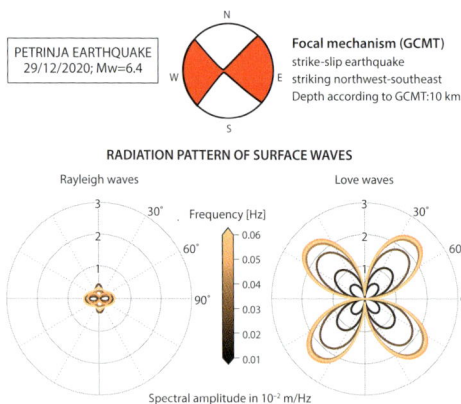

FIGURE 5.1. Top: Focal mechanism or the so-called beach ball of the Petrinja earthquake according to the Global Centroid Moment Tensor earthquake catalog (https://www.global-cmt.org/). The compressional motion for various distances and azimuths from the earthquake is projected onto the southern hemisphere of the imaginary sphere around the epicenter in red, and the dilatational motion is projected in white. In this way, four separate parts of the sphere were obtained. Bottom: Surface wave energy radiation diagram for various frequencies shown in various colors.

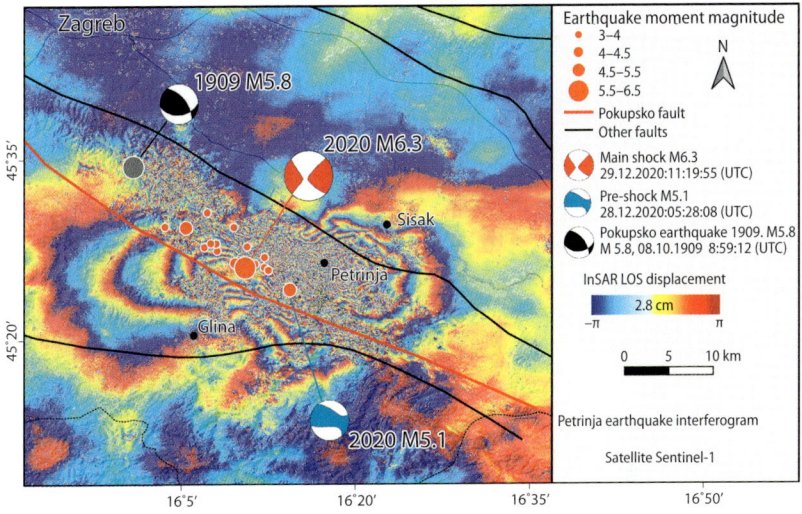

FIGURE 5.2. The wrapped interferogram of the Petrinja earthquake. The original interferogram using data from the Sentinel-1 satellite was made by Dr. Marin Govorčin from the Jet Propulsion Laboratory. The beach-balls of the preshock and the main Petrinja earthquakes were taken from the United States Geological Survey, and the beachball of the 1909 Pokupsko earthquake was taken from Herak et al. (2010). The faults are taken from Prelogović et al. (1989).

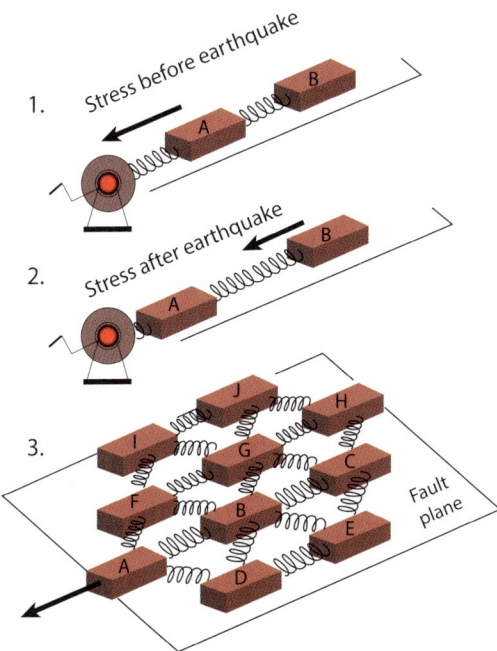

FIGURE 5.3. Schematic representation of system tension 1) before and 2) after the earthquake along a line. 3) Similar to 1) and 2), however, the stress distribution is shown in the plane representing the fault surface.

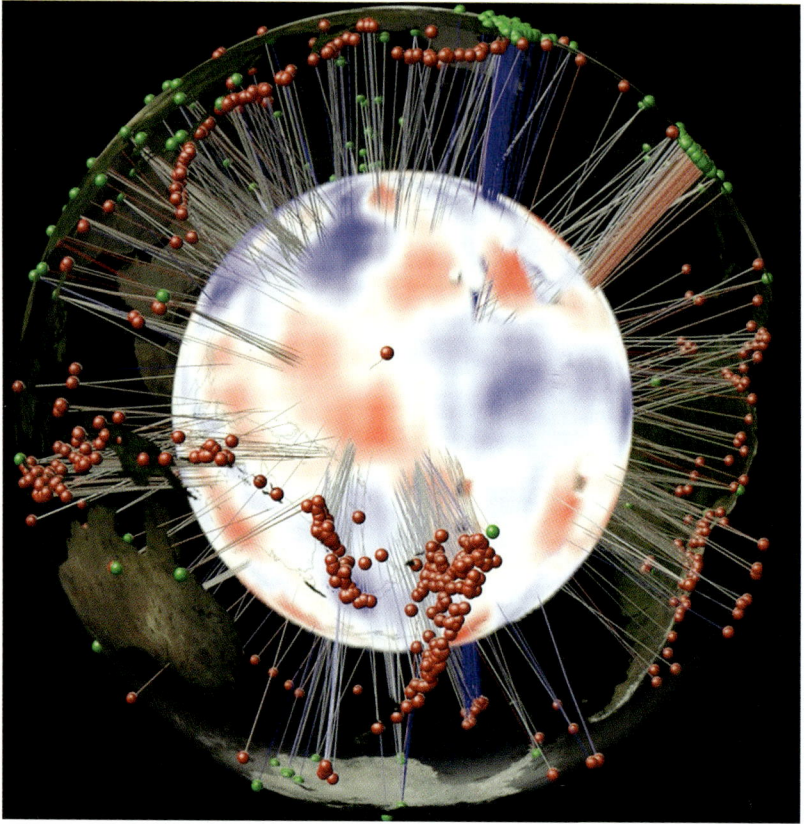

FIGURE 6.1. A tomographic map of the lowermost mantle projected to the core-mantle boundary. Red areas are slower, and blue areas are faster for the PKP waves than average. Ray paths of PKIKP waves from large earthquakes (their locations are shown by red spheres) through the Earth's interior to seismological stations (green spheres) in a three-dimensional view focusing on the Pacific. Note extensive seismogenic zones surrounding Australia, from the Indonesian earthquakes in the north to the prominent Vanuatu-Fiji-Tonga-Kermadec (in the clockwise direction) earthquakes in the northeast. The rays of seismic waves connecting the earthquakes with the seismic stations are shown in color, where red corresponds to slower and blue to faster-than-average values. The Earth's mantle is transparent for illustration purposes. The snapshot is from an animation created by Vizlab National Computational Infrastructure in collaboration with the author. The tomography was presented in the publication Tkalčić et al. (2002).

FIGURE 8.1. The redness of the Australian outback on the Oodnadatta track, near Marla, South Australia, about 235 km north of Coober Pedy. Our field vehicle in the evening sun under the rainbow.

FIGURE 8.2. Wild camels near the epicenter of the moment magnitude 6.1, 2016 Petermann Ranges earthquake, near the triple junction of the Northern Territory, South Australia and Western Australia borders.

FIGURE 8.3. The iconic Uluru (Mt. Ayers Rock) in the evening sun.

FIGURE 8.7. Termite mounds dominate the landscape at the Warramunga Seismic and Infrasound Facility, the Northern Territory, along one of the seismic array tracks.

FIGURE 8.8. At Karlu Karlu (Devils Marbles), about 100 km south of Tennant Creek, along the Stuart Highway. Professor Brian L. N. Kennett took the photo during my first Warramunga trip.

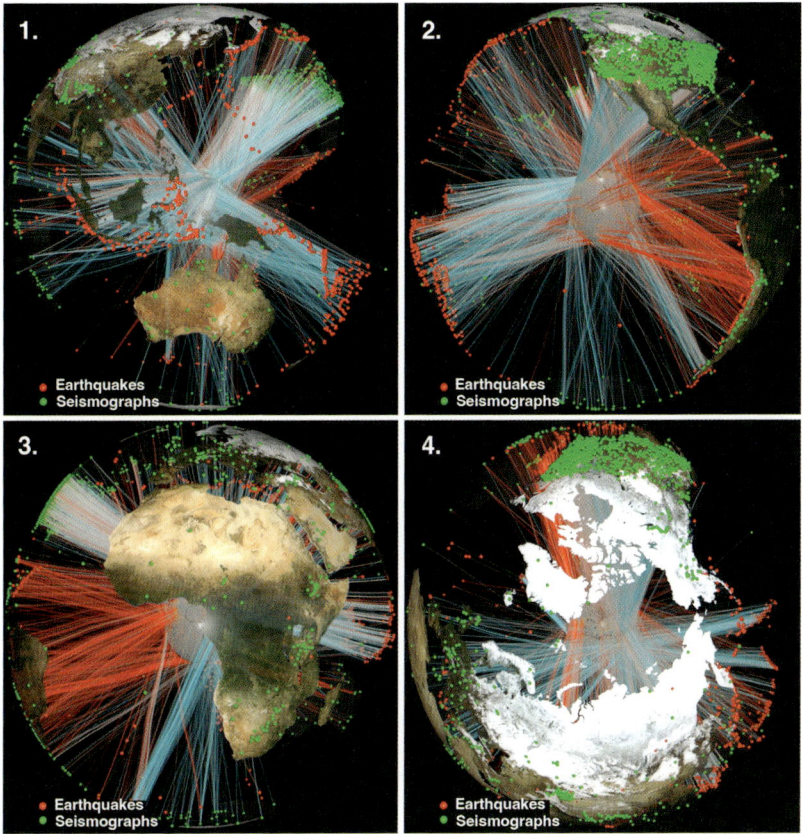

FIGURE 9.1. Ray paths of PKIKP waves through the Earth's interior from large earthquakes (their locations are shown by red spheres) to seismological stations (green spheres) in a three-dimensional view for four different perspectives: 1) Australia with surrounding seismic zones, 2) the Pacific Ocean with the 'ring of fire earthquakes, 3) Africa and the Mediterranean and 4) the Northern Hemisphere with a dense network of instruments in North America. The rays of seismic waves connecting the earthquakes with the seismic stations are shown in color, where red corresponds to slower and blue to faster-than-average values. The Earth's mantle is transparent for illustration purposes. The recordings were taken from an animation created by Vizlab National Computational Infrastructure in collaboration with the author, and the travel times of the waves passing through the Earth's interior shown in the illustration were measured by Dr. Lauren Waszek from James Cook University. We used the data to research the inner core, and several papers are in a review at the time this book is completed.

FIGURE 9.3. The Marine National Facility's RV Investigator in the Hobart harbor before our IN2020.V06 voyage in 2020.

FIGURE 9.4. Ocean bottom seismometers (OBSs) on the deck of RV Investigator before the IN2020.V06 voyage. The Australian pool of OBSs is in the front (yellow), while the China Academy of Science OBSs on loan for the project are in the background (dark orange).

FIGURE 9.6. A night deployment of an ocean bottom seismometer during the IN2020.V06 voyage.

Tonga, and Kermadec, all the way to New Zealand and the Macquarie tectonic complex in the south (see figure 6.1). Because of such a good distribution of earthquakes from all sides of the continent, it is possible to conduct excellent tomography. Thus, the first tomograms of sinking subduction zones in the Earth's mantle that went around the scientific world showed a section of the mantle just a little north of Australia. This became possible because my colleagues in the early 1990s carried and installed seismographs in the field across the Australian continent.[14] They recorded ground motions after significant earthquakes from around the world, and seismograms were then used to create the first tomographic images— figuratively speaking—of continental architecture, including the continental roots and their surroundings.

Australia's red interior looks more like the surface of Mars than Earth, not just from space but from the inside (fig. 8.1). When you are in the field, you are immersed in that surreal, almost mystical world of red vastness. Photoshop algorithms often behave unexpectedly because they misinterpret the red color of your photos. Sometimes, it's easy to forget that photos are created in moments when you are relaxed; however, fieldwork is often tiring and even painful. The air is hot and dusty, and unpaved roads act like icy surfaces for four-wheel-drive vehicles, with a much higher center of mass than standard vehicles. That's why the most significant hazard when you're out in the field in Australia isn't from crocodiles, spiders, or snakes but from your SUV. In addition to the relentless Sun, which

14. See, for example, the publications by Kennett and van der Hilst (1996), van der Hilst et al. (1998), and Debayle et al. (2005) in the bibliography. In recent times, Nick Rawlinson (see figure 9.2), now at the University of Cambridge, covered practically the entire southeast of Australia with a mobile network of seismographs.

FIGURE 8.1. The redness of the Australian Outback on the Oodnadatta track, near Marla, South Australia, about 235 km north of Coober Pedy. Our field vehicle in the evening sun under the rainbow.

burns so hot that you can make scrambled eggs on your vehicle's hood, you are sometimes surrounded by clouds of flies persistent in finding your mouth, eyes, and nostrils in search of protein. Those are not the happiest moments for photography when you hit a wild animal while driving or get stuck in the mud.

Although the Australian continent itself is far from the plate edges, this does not mean that earthquakes do not occur in its interior. For example, central Australia was hit by a strong earthquake with a magnitude of 6.0 in May 2016. The earthquake occurred in the Petermann Ranges, on the border of the Northern Territory and Western Australia, sandwiched between the Great Victoria Desert to the south, the Simpson Desert to the northeast, and the Great Sandy Desert to the northwest. The chain was formed about half a billion years ago, not so long ago for Australia, but enough to be older than most of the dinosaurs that once roamed these expanses. And therefore, whenever you are in the field, you can say that you are stepping on a large piece of fossil.

Immediately after the earthquake, drones searched the area
for any traces of faults on the surface. We installed six seismographs in the broader area around the fault to monitor aftershocks. That is a standard practice because much can be learned
about the mainshock and fault architecture from the spatial
distribution of aftershocks on a fault. About four months later,
we set out to collect them, moving clockwise in a curve about
1,200 kilometers long. We returned from the field in one piece,
avoiding many one-humped camels, which have been living in
the wild since they were imported from Afghanistan and India
in the 19th century for transport purposes (see figure 8.2).

Australia is a harsh continent that most people who live on
its edges do not have the opportunity to experience. I have had
it, and I am grateful that my life calling enabled it. I remember
my first fieldwork in Western Australia well. By area, it is the
second-largest federal state anywhere in the world, right behind
the Republic of Yakutia. Apart from this part of the continent
being far from larger human settlements, the oldest rocks on
Earth were also found there. They are dated to about 3.8 billion
years ago, before the beginning of plate tectonics. Indeed, the
Pilbara craton, a stable part of the continental lithosphere, has
never undergone subduction. The fossils of *Cyanobacteria* from
the Pilbara are the oldest direct evidence that life existed on
Earth about 3.5 billion years ago.

The expedition's goal was to pick up 20 seismographs installed about a year before I arrived in Australia.[15] They lay
patiently on the fossils of the Pilbara, far from ocean shores,

15. This deployment was installed by my colleague Anya Reading, who is now a
professor at the University of Tasmania. She has also recently installed a small array
in Antarctica as part of the project funded by the Australian Research Council we
work on together.

FIGURE 8.2. Wild camels near the epicenter of the moment magnitude 6.1, 2016 Petermann Ranges earthquake, near the triple junction of the Northern Territory, South Australia, and Western Australia borders.

people, and infrastructure, recording the tremors of the Earth from distant earthquakes, lightning strikes, dust devils, and the approach of the occasional curious dingo. The data allowed us to peer into the bowels of the Earth beneath the Pilbara. The two-week 6,000 kilometer off-road trip, starting and ending in Perth, was full of challenging conditions and tense moments.

The fantastic landscapes that opened on the horizon were riddled with tall termite mounds and *Spinifex* grass. Only Mt. Augustus managed to overshadow them with its appearance. Mt. Augustus, the largest monoclinal rock in the world, was formed by the bending of the Earth's layers due to tectonic forces. The age of Mt. Augustus is about a billion years. Erosion has consumed the sedimentary rocks at the top and exposed a massive piece of granite on the surface of the Earth, about 8 kilometers wide and 750 meters high above the sandy plain.

In the air, its iron minerals rusted and gradually turned red. The redness of the landscape and termite mounds, the white skeletons of kangaroos in cave openings, petroglyphs on ancient rocks, lonely tractors on dusty roads, the warm smell of the desert, the howling of dingoes, and the rise of the full moon over Mt. Augustus are just some of the images of this ancient, unfathomable world, which are profoundly and forever engraved in my memory.

However, unlike Mt. Augustus, which most people have never even heard of, Ayers Rock or Uluru is one of the first associations with Australia—its red heart (fig. 8.3). You will find a photo of Uluru hanging on the wall of almost every Australian embassy in the world or in the homes of many who remember their trip to the land "down under" with longing. It looks awe-inspiring, especially at dawn or before sunset. In 1993, a law was passed that officially allowed two names, Ayers Rock and Uluru, and since 2002, the national park has been called Uluru / Ayers Rock. In addition to rising from the base to a height of 348 meters, this stone island is estimated to descend into the interior of the Earth for at least another 2 kilometers. Uluru also has a "sister" named Kata Tjuta in the immediate vicinity (fig. 8.4).

The age of Uluru is estimated at 550 million years, at the time of the final phase of the formation of Gondwana, the southern supercontinent, from which Australia was later formed.[16] When Gondwana was formed by merging large boulders of the Earth's lithosphere or cratons, of which we have already mentioned the Pilbara, the Petermann mountain range was also developed, which was once as high as the Alps or the Himalayas. Sediments were deposited at the foot of those ancient mountains. Due to tectonic forces, the mountains continued their evolution and

16. See also the explanation in the glossary.

first reached below the ocean's surface, as well as a large part of the interior of Australia. A little later, between 450 and 300 million years ago, the mountain range "emerged" again, but because of erosion, which was more intense due to the lack of vegetation than elsewhere, today it probably looks slightly less impressive.

As a result of pressure and temperature changes, the sea surface sediments gradually formed very compact rocks. When Uluru and Kata Tjuta finally emerged from the softer rocks, the horizontal layers of sediments from which they were made were rotated almost perpendicular to the Earth's surface, which is how we see them today. As with Mt. Augustus, the red color on the surface of their rocks appeared as a result of the increased amount of iron in the rocks that oxidized in the air for hundreds of thousands of years; otherwise, it would have remained grey.

———

When I started exploring the Earth's inner core in 1997, I had no idea that my calling would one day lead me here and that I would spend a good portion of my time driving the Stuart Highway (see figure 8.5 for the main stops along the Stuart Highway). As you drive, you are sometimes jolted out of your thoughts by the motionless figure of a kangaroo or cow beside the road. Sometimes, it takes you a while to tell if it's a wild animal that has felt the vibrations of a speeding vehicle, or a giant termite mound. Aborigines occasionally dress large sculptures of termite mounds in worn clothes, so for a moment, they fool you into thinking you are not alone. This road, or, as the Australians have long called it, the "highway" is named after the Scottish explorer who was the first to survive the expedition to pass through the interior of Australia, from its southern part to the northern coast, in the sixth

FIGURE 8.3. The iconic Uluru (Mt. Ayers Rock) in the evening sun.

attempt, in 1862. The expedition was followed by efforts to establish a telegraph line and connection with Europe via Java, which then received a new telegraph cable.

The Stuart Highway largely follows the former Stuart route and telegraph line and cuts the continent in roughly two equal parts. It is about 3,000 kilometers long, from Port Augusta near Adelaide in the south to Darwin in the north. It passes first by lakes and lagoons, then through the most arid part of the continent, its red center, and finally through the savannah and tropics. When heading north on the road, the first stop for many is Coober Pedy, the so-called opal capital of the world, about 850 kilometers north of Adelaide. If you don't know it from the opals that were found there quite by accident in 1915, maybe some of you will recognize it from the filming locations of *Mad*

Max.[17] Many of the 2,000 souls who made it to Coober Pedy in search of a better life live underground because of the high temperatures, which often reach 50 degrees Celsius. The idea is not bad; since they were already digging to find gold and opals, why not turn it into a valuable and functional living space?

Scorching days and cold nights are typical for the desert climate, but these temperature fluctuations are almost nonexistent beneath the Earth's surface. This is how underground motels, shops, schools, and churches were built. Among the underground attractions are the Catholic church of St. Peter and Paul and the Orthodox Church, whose walls are covered with sculptures of saints carved into the rocks. They have become tourist attractions. You should not be surprised to hear familiar languages in some cafés and pubs. In my underground motel, I stared at something on the ceiling that looked like an upside-down umbrella and realized it was there to catch a few dust particles from the ventilation system. I remained somewhat taken aback by the excellent quality of the internet connection beneath the surface of the Earth.

On its northern side, the Stuart Highway begins in the city of Darwin in Australia's Northern Territory, which, with its area of almost 1.5 million square kilometers, ranks 11th on the list of the largest federal states in the world, right behind the Brazilian Amazon and Canadian Quebec. So, it is a vast expanse which stretches from the most arid part of the Australian continent on the border with South Australia, across the savanna and subtropical north, to the coasts of the Timor and Arafura seas and the Gulf of Carpentaria. Darwin, with its population of about 150,000 inhabitants, is not a candidate for some seriously

17. This film, set in dystopian near-future Australia, was directed by George Miller in 1979.

FIGURE 8.4. The fieldwork assistant, Armando Arcidiaco, fixes a flat tire on a track near Kata Tjuta (The Olgas).

remote territory, at least according to the definition of the Australian Government, according to which there are four categories of remoteness.

My destinations are often Tennant Creek and Warramunga, in the middle of the continent. You can reach them from Darwin after driving a thousand kilometers (fig. 8.5), but you may be attracted by one of the national parks on the way, for which you should turn east or west. We have already mentioned the termite mounds that turn into high castles and cathedrals toward the north (fig. 8.6). Their color corresponds here to the color of the soil, which, along with the sharp *Spinifex* grass and termite saliva and other secretions, is the basic structure of the termite mound. Some are up to 7 meters high, weighing 10 tons or more. Their interior is complex and interwoven with an

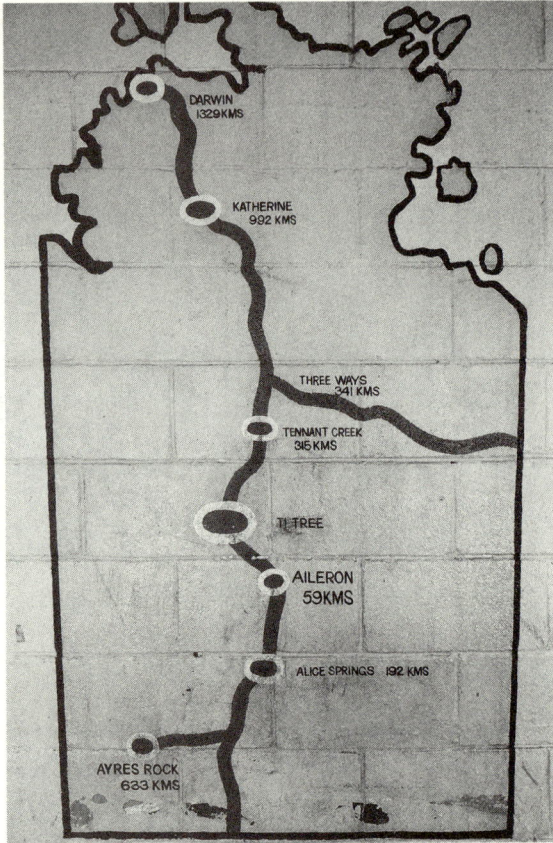

FIGURE 8.5. The northern section of Stuart Highway is displayed on a wall at the Ti Tree Road stop.

infrastructure of tunnels, some of which continue underground for 50 meters.

Thicker and shorter, narrower and taller, like giant human sculptures, figures of chubby friars and gaunt-faced saints with long beards dominate the landscape of the Northern Territory. Although the so-called magnetic termite mounds were initially thought to be built parallel to the magnetic field lines, it turned out that they were constructed according to the position of the Sun so that the Sun never directly illuminates the wider surface of the termite mound. That is why they seem like tombstones of some colossal and long-abandoned cemetery. In addition to termites, goannas, pythons, and other animals find refuge from the strong Sun in their cavities and hidden chambers.

While the American Civil War was winding down and ended in 1865, Frederick Henry Litchfield led a small group of European explorers and adventurers to the crocodile-infested Daly River in Australia. The expedition aimed to investigate whether there were conditions for settlement somewhere. In addition to magnetic termite mounds, they found Lost City—a sedimentary rock formation reminiscent of the buildings of an ancient civilization. Today, Litchfield National Park, in addition to its termite mounds and landscape (fig. 8.6), attracts visitors with its many hidden waterfalls and lakes where you can swim before the monsoons arrive. During the monsoons, the area becomes flooded by streams, and the lakes become accessible to crocodiles.[18]

But let's step for a moment now from western and central Australia's red expanses to those of New South Wales in the east of the continent. Don't be fooled by the fact that Sydney is the capital of New South Wales. Because if you thought we'd leave remoteness and red dust far behind in Western and South Australia or the Northern Territory, you'd be sorely mistaken.

One of the most remote parts of Australia I could visit during the seismological instrument installation campaigns was

18. My favorite swimming spot, of course, when there are no crocodiles, is Surprise Creek Falls, where the water has carved out rock pools on three natural terraces—a magical place to discover and worth the effort to get to, even when this includes a quick water crossing in an SUV across the Reynolds River, home to estuary crocodiles.

FIGURE 8.6. Cathedral termite mounds in Litchfield National Park, the Northern Territory, are among the tallest animal structures in the world.

New South Wales; namely, its northwestern part, near the border with South Australia, about 1,000 kilometers as the crow flies from Sydney. We made two separate expeditions in that northwestern corner of the state, on the edge of Sturt's and Strzelecki's Stony Deserts, and from the middle of the state toward its eastern coast, all the way to the fertile pastures and vineyards of Orange County, which somewhat resembles parts of Europe. In each of those two stages, we covered about 6,000 kilometers.

Speaking of Europeans, James Cook was probably the first to reach this area in 1770, sent to survey and chart the east coast of New Netherlands. New South Wales—as Cook called it— was politically founded as a penal colony of the British Kingdom in 1788. This was at a time when, after the American Revolution, the routes for the shipment of convicts to North America were closed. Without the allocation of convict labor to farmers and government projects such as road building and other infrastructure, the colonization of Australia would probably not have been possible. Convict labor was part of the sentence many in the New World served; however, New South Wales ceased to be a penal colony by 1852, well before it became part of the Federation of Australia in 1901.

I will remember the expeditions and fieldwork in this part of the continent for the period immediately after heavy rains when flora and fauna finally came to life after long periods of drought that preceded it. This was felt most during the drive because, in addition to the inevitable kangaroos and emus, many lizards, snakes, birds, grasshoppers, and other wild animals also perished under the wheels of our Landcruiser. Many people wonder if this suffering of animals can be avoided. It's possible if you don't go this way at all. Truthfully, when you have thousands of wild animals in front of you in a stretch of 6,000 kilometers, close encounters are impossible to avoid. You drive on narrow field roads, through tall grass, along abandoned riverbeds, and along kilometers of wire fences, and when a wild animal jumps out of nowhere in front of you, instinctive braking can cost you your life.

On the one hand, SUVs have a higher center of mass than ordinary cars, and when braking hard, they are easy to drift and overturn. On the other hand, emus and kangaroos have worse peripheral vision than cats or dogs. If you don't see them in time, the best thing you can do when the emu starts to run a slalom in front of you is to brake with your engine by switching from a higher to a lower gear and try to maintain the direction of movement.

The Stony Desert in the western part of New South Wales stands out for the phenomenon of densely interwoven paths of stone pebbles cemented by the ravages of time into a mosaic of desert soil. It is believed that these tracks were formed by the gradual loss of sand and dust from the upper layers due to wind and precipitation, and the larger pebbles gradually fit into the puzzle due to the microseismic shaking of the Earth, which, like a colossal sieve, constantly shook and rearranged the pebbles. Once the mosaic had formed to the point where all the pieces were in place, and there was no more free space between them, the desert took on its final appearance. This phenomenon is also present on the outskirts of other large world deserts, such as the Sahara and the Mojave Deserts.

As we moved eastward, the grass grew taller, and dense swarms of locusts, which we plunged into like black clouds, became our daily reality. We had to stop occasionally and briefly to check the engine radiator, where tall grass used to get stuck. Every time, the characteristic smell of roasting that came from the front of the vehicle and under the hood would briefly take me back to my childhood and the countless times I passed by a rotisserie on the way to school. Eucalyptus, European trees, and Orange County's first vineyards gradually replaced grasses and undergrowth. When I saw Zinfandel on the list of one of the wineries, I knew we were almost home.

———

As I get closer to Tennant Creek, more and more signs are on the road (see figure 8.6 for the location of Tennant Creek along the Stuart Highway). You get used to seeing a few advertisements for mines, a car repair shop, an Aboriginal gallery, a motel, and a war veterans club with a restaurant on approaching small towns when you've been in the field up and down the Australian continent. Tennant Creek was a mecca for adventurers, eccentrics, and gold diggers. One of them, Jack Noble, appears in a photo on the wall of the Memorial Club established during the gold rush of the 1930s. But it should be said that the settlement next to the telegraph station has existed since the 1880s and that Jurnkkurakurr, another name for Tennant Creek, is where the Warumungu tribe has lived for thousands of years.

For all the reasons mentioned above, fieldwork on the proliferation of seismographs in inaccessible parts of the continent became part of my scientific calling, on average, once a year. However, I have been coming to Tennant Creek for years for another reason.[19] At the center of the turbulent geopolitical events of the Cold War, the establishment of the L-shaped Warramunga array of 20 seismographs was anything but an ordinary story (fig. 8.7). Warramunga is far from it all, on Warumungu tribal land. The Research School of Earth Sciences at the Australian National University began administering it shortly after it was established. And after 50 years, we still manage Warramunga on behalf of the United Nations Comprehensive Test Ban Treaty Organization (CTBTO) and the Australian Government. But why do we do it at all, and what does it have to do with seismology and my scientific interests?

We briefly looked at the fundamental difference between tectonic earthquakes and explosions in the discussion about earthquakes. The movement of tectonic units along faults causes tectonic earthquakes; however, there are also volcanic earthquakes, in which the sudden movement or expansion of magma, liquids, or gases under pressure can cause seismic waves. Their characteristics are significantly different from the characteristics of tectonic earthquakes. Meteorite impacts,

19. Coming to the Australian National University, I took over, as director, management of the Warramunga Seismic and Infrasound Facility.

FIGURE 8.7. Termite mounds dominate the landscape at the Warramunga Seismic and Infrasound Facility, the Northern Territory, along one of the seismic array tracks.

landslides, glacial earthquakes, tornadoes, cyclones, mine blasts, chemical explosions in factories and warehouses, and many other phenomena can be detected on seismographs. And, of course, there are also nuclear explosions. It turns out that modern seismological methods can be used as forensic methods to distinguish an earthquake from a bomb. This is why seismology is one of the primary core competencies inherent in discriminating explosions from earthquakes.

Even if you had a lot of earthquakes under your belt, you might never be able to tell for sure whether it was an earthquake or an explosion based on your senses alone; however, the Warramunga facility—a modern array of seismic and infrasound devices—can greatly surpass human senses. It consists of 24 seismic broadband and 8 infrasound elements. When you have

24 instruments arranged in a carefully designed configuration in a territory far from the coast and any significant human activity, you have something like—figuratively speaking—a giant telescope pointed at the center of the Earth. As for the infrasonic waves, they travel from the blasts through the atmosphere as perturbations in the air pressure and get recorded by highly sensitive microbarometers. Therefore, Warramunga is the primary station for seismic and infrasound monitoring in the International Monitoring Instrument Network. The data recorded in the Australian savanna are sent via satellite link to Vienna, the center of the International Monitoring System.

The difference between the waves generated by bombs and those caused by earthquakes is that during an explosion, mainly compression waves are generated. In contrast, during earthquakes, on seismograms, we clearly distinguish between compressional and shear waves. Generally speaking, this difference can be used for forensic purposes to identify hidden nuclear explosions and differentiate them from natural events. Therefore, it should not be surprising that in the wake of the Cold War, seismology flourished as one of the key scientific methods for "eavesdropping" on rogue states that were developing nuclear programs. This ultimately improved the quality of seismographs and contributed to the development of broadband instruments, improved versions of which we still use today.

The United States led the way with the number of tests (1,030), followed by the Soviet Union (715), then France (210), China (45), Great Britain (45), India (3), and Pakistan (2). In addition to development of instruments, the Cold War also resulted in installation of seismographs in remote and inaccessible parts of the continents. For example, it is not difficult to imagine why the United States has installed numerous networks in Alaska. The Warramunga array of seismographs and infrasound sensors in the center of the Australian continent was also established at that time. The quality of the Warramunga observatory, due primarily to its distance from the coast of the ocean and big cities, can only be matched by the observatory of the same purpose—Yellowknife in the Canadian Northwest Territory, and that only in the Canadian winter, because it gets noisy in the summer due to ice melting and cracking on the nearby lake, which is undesirable for listening to distant earthquakes and explosions.

Aside from the essential difference between explosions and earthquakes that manifests in different types of seismic waves, many will be interested in how much energy is released in an earthquake compared with some of the nuclear explosions. To clarify this, it is first necessary to mention that the energy value in physics is expressed in joules. Still, it can also be expressed in other units, such as calories. For most of you, calories are probably intuitively closer.

The average person needs about 2,500 calories a day. So, a little less than 1 million calories are enough for one person in a year. A thousand people need about 1 billion calories in a year. That is roughly equivalent to the energy released in the explosion of one ton of trinitrotoluene (TNT), often used as the reference unit of explosives. If you are more familiar with calculations of consumed electricity, 1 ton of TNT is approximately equivalent to 1,000 kilowatt-hours, roughly equivalent to a large family's monthly consumption.

As much as 1 ton of TNT may seem like a lot to you—how could it not be when it is enough energy for a thousand people a year—it is only a tiny part of the energy value of the explosives that were tested during the Cold War. Let's take an example of the biggest bomb in history. The Soviet Tsar Bomba of 1961 was equivalent to releasing the energy of about 100 megatons of TNT, or 100 million times the energy of one ton of TNT. Now, to put the energy released during more significant volcanic eruptions or earthquakes in the context of explosions, the Krakatoa eruption of 1883 was twice as energetic as the Tsar bomb, so 200 megatons. But the Sumatra Earthquake of 2004, with a magnitude of 9.1, whose tsunami ravaged the countries along the Indian Ocean, released energy equivalent to the explosives of as much as 10 million megatons of TNT. Hence, it is a number with 13 zeros or, in jargon, 13 orders of magnitude more than 1 ton of TNT. You'll have to admit that these orders of magnitude make your head spin, but for the curious, just this: if you're wondering how much energy that is compared with the asteroid impact that caused the extinction of the dinosaurs 65 million years ago, that asteroid impact released about 100 to 1,000 times greater energy than the largest recorded earthquake.[20]

Be that as it may, the Soviets, Americans, and British stopped nuclear tests in the early 1990s, the French and Chinese in 1996, and the Indians and Pakistanis in 1998. More recently, North Korea has conducted six experiments, the last of which was in 2017. In the meantime, from the end of the 1990s to today, the number of seismographs in the world has increased

20. The discovery of the Chicxulub crater, caused by the impact of a giant asteroid that caused the extinction of dinosaurs and a large number of other living species at the end of the Cretaceous period in the Mesozoic era and the beginning of the Paleogene period in the Cenozoic era, was beautifully described by Walter Alvarez in his book *T. rex and the Crater of Doom* (1997). Walter Alvarez was on my doctoral committee at Berkeley. With his father, physicist and Nobel laureate Luis Alvarez, he hypothesized that the extinction of the dinosaurs was caused by the impact of a giant asteroid. There was no trace of the crater for a long time, but it was finally found in the early 1990s near the Yucatán Peninsula in Mexico.

by two to three orders of magnitude (100 to 1,000 times). Of course, I did not want to say that the proliferation of seismological instruments in the world occurred solely because of the Cold War and efforts against the increase in nuclear weapons. In fact, nuclear nonproliferation forensic seismology is a topic unto itself. But the truth is that investments were made in measuring instruments and increasing their density on the planet's surface. Some of the most important discoveries about the Earth's deep interior came solely because of the existence of Warramunga and other high-quality facilities and instruments. Sitting in rusted earth, inside the buried concrete pipes, hidden from inquisitive eyes, modern seismographs can "sense" a fly flying over the tall grass of the savanna.

I consider myself privileged that, as someone from academia, I have been able to participate in the worldwide efforts to monitor and prevent the spread of nuclear explosions for the past 17 years.[21] I firmly believe in recognizing the uniqueness of facilities like Warramunga, scattered worldwide, and their unique needs depending on conditions. It is easy to make a slight misjudgment from the Vienna office. You realize this, for example, when, after major infrastructure reconstruction, you find two large goannas wrapped around each seismometer, which have decided to spend their savanna nights in firm embrace of our equipment. In retrospect, focusing only on the science of earthquakes and explosions would have been much easier. If you ever thought that scientists were strong-willed, visit the offices in the long corridors of the United Nations building in Vienna!

21. I perform this duty by managing Warramunga from my office in Canberra, going to the field and traveling to Vienna to the Procurement of the Comprehensive Test Ban Treaty Organization (CTBTO) at the United Nations, where I interact with diplomats and accountants. I stand for the future of Warramunga and its role.

My annual tour of Warramunga usually falls in the late monsoons, first with a flight to Alice Springs and then a 500 kilometer drive north to Tennant Creek. The path across the Tropic of Capricorn leads you to the savanna belt, a kind of entrance to the subtropical north of the continent, until you eventually reach Tennant Creek. You then drive to the Warramunga facility, an approximately 50 kilometer stretch from Tennant Creek, just like the two technicians do each day. Every year, they endure extreme heat, dust, flies, fires, floods, thunderstorms, scorpions, snakes, goannas, and many other natural elements that would be impossible to list here. I tried to get there as quickly as possible during emergencies and in other, calmer times. I was their support from afar in some of the most dramatic moments of fire and floods.[22]

And what happened to Warramunga and me all these years? With each visit, I became happier, in the touch of the warm monsoon wind on my face and skin and the breath of the hot savanna before the long-awaited rains, in the wavy grass, among the red rocks, with the clop of wild horses, and the whistling of distant dust devils, scared in moments of nature's fury and composed in times of silence. Slowly but surely, she seduced me with her quiet and graceful beauty beyond every word and image.

———

The night at Tennant Creek is warm and humid. A small town sleeps with a legacy of pioneers, adventurers, eccentrics, and miners, a not-so-distant past of living far from everything, a

22. Many of my colleagues at the Australian National University and former and current employees at Warramunga deserve praise—without their sacrifice and support, the functioning and maintenance of Warramunga would be impossible.

FIGURE 8.8. At Karlu Karlu (Devils Marbles), about 100 km south of
Tennant Creek, along the Stuart Highway. Brian L. N. Kennett took
the photo during my first Warramunga trip.

scarcity of water and necessities, and the fates that marked this
"wild west." I can't wait for the alarm at 4:30 because of the
adrenaline. I turn it off in time so as not to wake the other guests
in the motel made of wooden trailers in the shade of palm trees
and other plants in a subtropical oasis. It's still pitch dark outside. No one on the road from the opposite direction in the first
100 kilometers. Just an occasional pair of wild animal eyes glint-
ing out of the darkness. From the satellite, I am again just a tiny
bright dot slowly moving south in the continent's center.

A bright strip of the coming day is born on the eastern horizon. I always plan my way to Alice to be alone at the Devil's
Marbles[23] before dawn to wait for sunrise (fig. 8.8). The place

23. In English, Devil's Marbles; in the native language, *Karlu Karlu*.

looks mystical at dawn, like the Stonehenge of the Southern Hemisphere. I have just enough time to take a few photos and go on my way to Alice.

"Watch out for the rain when it comes down, darl," a middle-aged lady at the Ti Tree roadside stand kindly says, handing me a giant plastic cup of black coffee. I think briefly about the size of the cup, which probably all the drivers of road trains— trailers 53.5 meters long—consider the most normal. I am escorted from the gravel parking lot by the smell of diesel and a beautiful peacock that has been reigning at this stop along the Stuart Highway for years. As I approach Alice, heavy raindrops slowly but surely wash the red dust off the windshield. I think about how effortlessly I can catch a plane to Canberra, leaving from the airport via Melbourne once a day.

9

At the bottom
of the ocean

"There is no law south of 40 degrees latitude. And south of
50 degrees, there is no God."

—OLD SAILOR SAYING

The prow of the research vessel (RV) *Investigator* makes its way
through the turbulent, murky waters of the Southern Ocean,
leaving behind the oblong Macquarie Island. Its steep slopes
look even more spectacular, dressed in white from the recent
snowstorm. The untrained eye would miss the studded black
dots at its base, right next to the ocean, blending into the shore-
line. There is no way to recognize the colonies of king penguins,
which, together with other species on the island, number more
than a million.

A strong, westerly wind hits surfaces in its path. The protective raincoat I received before our voyage vibrates under its
force and drowns out the sounds of the waves lapping the deck.
A lone albatross levitating from a height calmly observes this

white colossus, the pride of the Australian scientific fleet, on its slow motion northward. With its widespread, almost motionless wings traveling along, it appears to be connected to the ship by some invisible pole.

Moving away from the subantarctic waters, I think about everything we've been through in recent months. Contrary to the sea mermaids who knew how to enchant sailors with their beauty and voices, the Nereids—daughters of Doris and Nereus, the eldest son of Gaia, helped the unfortunate sailors who found themselves on a stormy sea. Those good nymphs of the ocean and some extra time helped us to complete the goals we had set for ourselves months and years ago and set sail with optimism to return to Tasmania. Or how else to explain that after weeks of pure agony and a race against time, we were "gifted" the entire remaining week of favorable weather conditions for sonar imaging of the ocean floor and lowering the remaining seismographs into its depths? Whatever divine intervention it was, I am happy that we are returning home after 27 seismographs have been successfully lowered to the ocean floor around Macquarie Island.

We still have another 810 nautical miles north to Hobart, and with some luck, we'll get there in 75 to 80 hours. I don't even have to say how much I look forward to the solid ground under my feet after almost a month of balancing on "sea legs" through the corridors of RV *Investigator*. Undoubtedly, all members of our scientific team are looking forward to the same. But why did we even dare to embark on this voyage and subject ourselves to the mercy of the Southern Ocean? After the demanding conditions in the interior of the Australian continent, why this extreme geophysical expedition? Why did we have to place seismographs at the bottom of the ocean? To answer these and similar questions, let's take a few steps back to the discussion about the coverage of the planet by large earthquakes and seismographs.

The current version of the global seismograph network is easy to find on one of the maps available on the internet. But to visualize all the world's seismographs and earthquakes in a slightly different way, imagine their locations on a transparent globe that we can rotate in our minds around a vertical or horizontal axis (fig. 9.1). Let's focus on only about 20 years of earthquakes greater than moment magnitude 5.8. We mark the locations of large earthquakes with small red dots and the locations of permanent seismological stations that record seismic waves with green ones. Let's simplify the wave paths with thin curves connecting the red and green dots through the globe's interior, but let's do this only for the waves passing through the Earth's inner core. Now, pay attention to a few facts.

First, if you could watch the entire rotation of the Earth around its vertical or horizontal axis, it would surely become apparent that certain parts of the world are covered by earthquakes and seismological stations perfectly, while others are eerily empty. Of course, you would immediately notice large earthquakes along the edges of the tectonic plates. The Tonga-Fiji-Kermadec subduction zone northeast and east of Australia would be particularly striking, and you would see the densely interwoven locations of the world's largest and deepest earthquakes there (fig. 9.1; top left panel).

You would immediately notice the dense network of instruments covering Japan (recall figure 2.1) and a series of earthquakes in the Aleutian Islands, connecting the Japanese islands with the west coast of America like a suspension bridge. If the globe in your mind turned from west to east, just as the Earth itself rotates, you would also see several significant earthquakes in the Himalayan and Mediterranean areas. The African continent would gape almost empty, with only a few green dots. Finally, you would spot a series of earthquakes off the west coast of the United States and a dense network of seismographs—the USArray—which has completely covered the territory of the United States of America.[1]

Second, for these types of seismic waves to pass through the Earth's inner core, it is necessary to have an earthquake and a seismological station on almost opposite sides of the world. The station locations would extend from North Africa to western Europe for earthquakes from Tonga and Fiji. The antipodes would be in Southwest Asia for earthquakes from Chile or Argentina. We could find existing seismological stations for both the first and second groups of earthquakes. But for some other earthquake locations, finding an antipodal station on the continental mass would be impossible. It would be somewhere in the ocean. The story is almost identical for any type of seismic wave we use to peer into the interior of a planet—many valuable locations for seismographs would be in one of the oceans.

Our planet appears blue from space,[2] so the source-receiver geometry with many desired seismograph locations ending up somewhere in the ocean should not surprise us. But it leads us to another realization: placing a seismograph on the bottom of the ocean would be a precious technological achievement, without which it would be impossible to progress, especially if we consider that we cannot influence the locations of major

1. The typical distance between two seismographs of this network is 70 kilometers (figure 9.1; panels on the right).

2. The Earth is predominantly blue from space because most of its surface is covered by oceans. Ocean water most efficiently absorbs ultraviolet, infrared, and red radiation, and the deeper it is, the more it gradually absorbs other wavelengths of visible light. That is why shades of blue can be seen from space, of which dark blue corresponds to the deepest parts of the ocean.

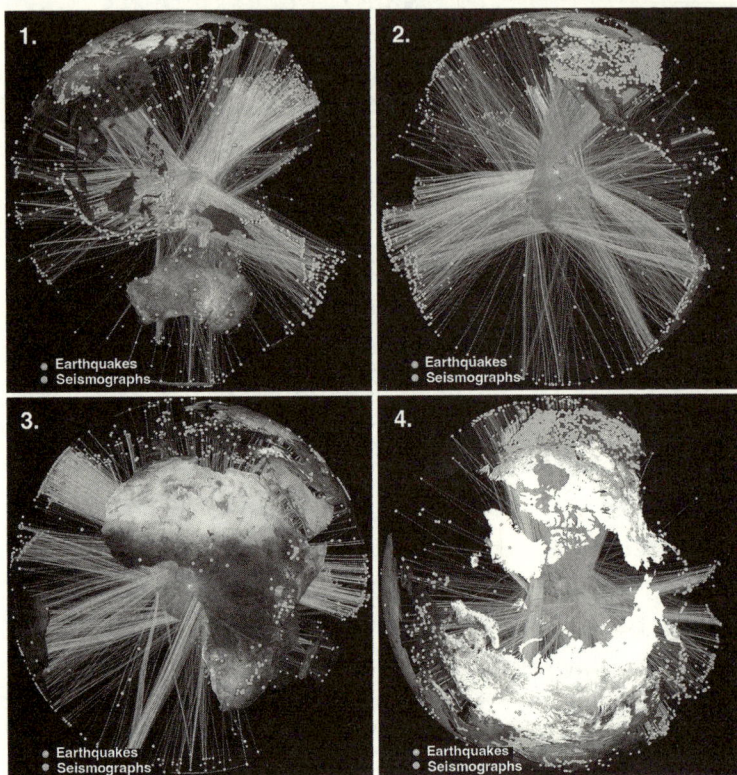

FIGURE 9.1. Ray paths of PKIKP waves through the Earth's interior from large earthquakes (locations shown by red spheres) to seismological stations (green spheres) in a three-dimensional view of four different perspectives: (1) Australia with surrounding seismic zones, (2) the Pacific Ocean with the Ring of Fire earthquakes, (3) Africa and the Mediterranean, and (4) the Northern Hemisphere with a dense network of instruments in North America. The rays of seismic waves connecting the earthquakes with the seismic stations are shown in color, where red corresponds to slower and blue to faster-than-average values. The Earth's mantle is transparent for illustration purposes. The recordings were taken from an animation created by Vizlab National Computational Infrastructure in collaboration with the author, and the travel times of the waves passing through the Earth's interior shown in the illustration were measured by Dr. Lauren Waszek from James Cook University. We used the data to research the inner core, and several papers are in review at the time this book is completed.

earthquakes. They are mainly on the edges of tectonic plates. And that's why we need seismographs at the bottom of the ocean to take a fair image of the Earth's interior.

————

They say life writes the best stories, and they're probably right. When I visited Tasmania for the first time about 15 years ago, I was not at all impressed with myself because I realized how little I knew about its history, the beauty of the coasts and hills, the Tamar River valley, and the secrets eternally written in the white stone blocks of the houses and streets of its picturesque small towns.

Fast-forward the film more than a decade, and I find myself in quarantine, locked in a hotel room near Salamanca,[3] thinking about the days ahead. The room is so tiny that it feels like I can touch all its walls if I spread my arms wide enough. I don't want to think about the view through the sealed window blocked by the parallel wall on the other side of the hotel, separated by only a few meters of air. I don't even think about the smell of the underground garage, where I stretch my legs every day for half an hour, with a mask over my mouth and nose and wearing surgical gloves. I'm not complaining because a bottle of Tasmanian champagne is just arriving from the reception, courtesy of my colleague, Mike Coffin, an old sea wolf from the University of Tasmania, with whom I'll be leaving in two weeks for my first big voyage aboard the main Australian research ship, RV *Investigator*.

In trying to bring Australia closer to you as a continent, I have often mentioned its Outback, which probably makes up

3. A famous part of Hobart, close to the harbor.

between 80% and 85% of the continent's surface. On the one hand, it is an antithesis to those who harbor romantic visions of Australia, because when Europeans visit Sydney or Melbourne and say they have experienced Australia, that should be taken with a grain of salt. On the other hand, lest the sentiments that Australia is just a harsh continent—practically, a fossil you walk on, dry as powder and red with rusted iron on the surface and inhabited by wild, exotic animals that can move you to the other world with one stroke—it's time to stay for a while in Tasmania, the starting and ending point of the expedition to the Southern Ocean, which has approached.

So, let's move even farther south from the green coastal belt of Australia to a small island some 250 kilometers south of the main continental mass. By the way, it should be said that the area of this "small island" with the surrounding islands is about 68,000 square kilometers, so it is much larger than the area of many European countries.

My first encounter with Tasmania was related to a network of seismographs installed mainly in the eastern, somewhat more accessible part of the island. We traveled along and across for several thousand kilometers. As a result of the experiment, we published tomographic images of the Tasmanian subsurface using the microseismic noise tomography method,[4] for which Tasmania is ideal because the Southern Ocean surrounds it on the southwest side and the Tasman Sea on the northeast. Namely, the constant interaction of the solid earth with the ocean mass creates microseismic noise. Tomographic images of underground architecture can be obtained from this noise by exploiting the way it spreads over the surface and underground, similar

---

4. See Young et al. (2011) in the bibliography.

to the tomography we have already described.[5] The geological structure and history of Tasmania and its connection with Australia are still the subjects of research and discovery.

The first European who arrived in this area was not a Briton but a Dutchman, Abel Tasman, in 1642. It is still a bit unclear why the Dutch left Australia to the British, but according to some stories, they first sailed around Australia from the "wrong" side, west and south, and found Indonesia more attractive. Be that as it may, Tasmania was a haven for sealers and whalers for a long time. When Great Britain appropriated it, it became a penal colony at the beginning of the 19th century. The island was long called Van Diemen's Land after the company governor who first explored it. Aboriginal people have lived here for the past 40,000 years. When the island became a British colony, there were about 10,000 of them, but after only 30 years, only a few hundred survived.

Staying in Tasmania was an excellent opportunity to learn about the history of this part of the world, see how people live, and explore the wines of a cool climate region determined by the surrounding ocean, which compete with those from New Zealand.[6] Our seismographs were located mainly in orchards and vineyards, and the plots of land there are much smaller than those in mainland Australia (fig. 9.2). I must admit that when I lived in California, my association with the subject of Tasmania was either the Tasmanian devil or the apple orchards from the Australian comedy *Young Einstein*. And indeed, the island abounds with both. In fact, if you're wondering, Tasmania was a significant exporter of apples for many years, mainly because they managed

5. See chapter 6.

6. Of the wine varieties, there are mostly Pinot Gris, Sauvignon Blanc, sparkling wine, and Riesling, with much more subtle aromas and flavors than the New Zealand equivalents, while Pinot Noir dominates the red wines.

FIGURE 9.2. With Nick Rawlinson and a herd of sheep, maintaining a temporary seismic network in Tasmania. The white box is a recorder that remains above the ground and must be protected from the animals, usually by wrapping it in a tarp and covering it with dry tree branches and other vegetation. The sensor is buried. Our fieldwork assistant, Tony Percival, took the photo.

to avoid disease. And speaking of diseases, Tasmanian devils almost died out due to a mysterious tumor of the face and head, which began to spread uncontrollably in the 1990s.

The main reason for our imminent departure from Hobart is its location; it is the center of MNF,[7] which administers maritime activities from the home port of the RV *Investigator* (fig. 9.3). The vessel is 94 meters long, 18.5 meters wide, and 10 stories high, with a capacity of 60 crew members, including scientists. Its cruising speed is 11 knots, and it has the ability to

7. Marine National Facility.

survive 60 days on the ocean with its fuel reserves. For those interested in the details, its gross tonnage is 6,802 tons. In addition to its impressive appearance, we were enchanted by its modern self-stabilizing system, which, by transferring water through huge reservoirs deep in the ship's bilge, establishes balance and counteracts large waves. That was to protect us from extreme rolling in the stormy waters of the Southern Ocean. However, apart from the excellent first impressions of the ship, we were also impressed by the professionalism of its crew. Graduate students and postdocs enjoyed boarding, comfortable cabins, and amazing food from the ship's chefs, which we all quickly got used to in the days after coming out of quarantine and just before sailing. Unfortunately, for many, it was also the last time they enjoyed food before returning from sailing.[8]

An essential part of the logistics was to deliver to Hobart 29 specialized seismographs for monitoring the ocean floor.[9] We managed to gather two groups of instruments: 17 with a design in the form of yellow rectangular prisms (fig. 9.4) and 12 in the form of orange-red spheres. The market value of each device was approximately AUS$200,000. To give an idea of their weight, each instrument, together with the heavy concrete weights that drag them to the ocean floor, weighs 250 kilograms.

Before departure, they had to be tied to the ship's deck to protect them from harm (fig. 9.4). And even before that, they had to be serviced and their batteries charged to survive on the ocean floor for about a year. The sensors of these specialized seismographs are sensitive to wave periods between tens of seconds at one end and around tenths or hundredths of a second at the other end of the seismic wave spectrum. Submarine seismographs are also equipped with sophisticated digitizers,

8. Because of seasickness.
9. We managed to collect that many instruments in functional condition.

FIGURE 9.3. The Marine National Facility's RV *Investigator* in the Hobart harbor before our IN2020.V06 voyage in 2020.

several types of batteries and discharge mechanisms, and weights. In addition to the three components of the seismometer for describing motion in all three dimensions of space, they also have hydrophones, sensitive to frequencies of only a few to several tens of hertz, which means that they can record the voices of whales inaudible to the human ear![10]

———

As we leave the Hobart harbor with a long and loud greeting from the ship's siren, which cuts the air with the power of a giant

10. Let's recall the discussion about the sounds of earthquakes in chapter 4.

FIGURE 9.4. Ocean bottom seismometers (OBSs) on the deck of RV *Investigator* before the IN2020.V06 voyage. The Australian pool of OBSs is in the front (yellow), while the Chinese Academy of Sciences OBSs on loan for the project are in the background (dark orange).

horn, a lump forms in the throat, and a watery eye creeps in. About 80 hours of sailing separate us from the nearest point of our destination. It is the northeast corner of a rectangular piece of the ocean, halfway between Tasmania and Antarctica, with an area of about 1,000 kilometers. We are supposed to lower 29 seismographs to the ocean floor in predetermined locations.

We spend the first night tumultuously, both literally and figuratively. Namely, sailing across the "roaring forties," as sailors call the belt of the 40th to the 50th degree of southern latitude, we rightly wondered what would happen when we got closer to the "furious fifties." They got their name from raging waters and the destruction they were known to cause back in the days when the first trade routes from west to east were being established. In fact, it can be said that these latitudes shaped trade routes.

The warm air, which rises from the equator and flows meridionally around the Earth toward the poles, descends back toward the Earth's surface about 30 degrees from the equator, forming a giant Hadley cell, one in each hemisphere. The air then continues its path across the surface toward the poles until it rises again at about 60 degrees latitude, forming smaller Ferrel cells. In the Southern Hemisphere, the air between 30 and 60 degrees latitude slowly turns eastward[11] due to the Earth's rotation. This wind is mighty in the Southern Hemisphere because, at these latitudes, no continental mass presents an obstacle. This uninterrupted passage contributes to creating the strongest zonal circulation on the planet, both the wind and the ocean water mass that the wind moves. This is also called the Antarctic Circumpolar Current, with a typical speed of about 4 meters per second. Translated into water mass transport language, this amounts to more than 100 million cubic meters of water per second or 100 *sverdrups*.[12]

With each new hour, and especially after the transition from the "roaring" to the "furious" belt of latitudes, it becomes clear that it will not be easy for us. The wind and the ocean do not let up, as if they never intend to stop. The purpose of a piece of chain under the office chair in my cabin is also becoming crystal clear to me now. Securing the chair to the floor, I reminisce about the training and health checks we underwent and think

11. In meteorology, winds are named according to the direction from which they blow, so this wind is also called westerly or westerly wind.
12. After Herald Ulrik Sverdrup, Norwegian oceanographer and meteorologist.

this experience will overshadow everything we learned before sailing. I wonder about the others; how are the other members of my research team coping with the situation?

Unfortunately, it soon turns out that seasickness has decimated the scientific part of the crew, who until then could be seen together during lunch and dinner. The food is replaced by nausea pills and a stainless steel thermos for water in the shape of a long submarine that everyone receives as a gift. In conversation with the ship's professional crew, I learn that seasickness is not something that disappears after you have gained experience and had a good number of voyages. In people sensitive to the ship's motion, it occurs every time, but it helps once you get used to it and know what awaits you. I think each of us has a different sensibility and adjustment period, and I hope everyone will be ready when we reach our destination.

We are divided into two shifts of 12 hours each to make the most of the expensive ship time made available to us through the competitive process. Twenty-nine instruments needed to be installed on the ocean floor one after the other at a given time, regardless of the time of day or night (fig. 9.5, 9.6). What needs to be done before we start lowering each of the seismographs to the bottom of the ocean is to map it; that is, as it is said in the jargon, we need to create a bathymetry—a map of the ocean depths. Since we do not have accurate information about the bathymetry of the ocean floor for this part of the planet, we will spend several days in each of the four quadrants of our drawn rectangle, probing the terrain with broadband sonar.

The ship is equipped with several types of sonar. An accurate ocean floor map will become available to us on the fly. It will allow us to select locations as close as possible to those

FIGURE 9.5. Deployment of an ocean bottom seismometer during the IN2020.V06 voyage.

we originally planned. Namely, we know that Macquarie Island rises as part of a ridge above the ocean surface, about 6,000 meters above the ocean floor. Still, we do not know precisely how the sediments and solid rocks are distributed on the bottom.

FIGURE 9.6. A night deployment of an ocean bottom seismometer during the IN2020.V06 voyage.

Since the sounding direction parallels the Macquarie Ridge, the westerlies will blow directly from the side and constantly tilt the ship. Once the maps are ready and the desired location is confirmed, the entire operation of arriving at the site, preparing and lowering the seismograph to the bottom, and determining its location using the triangulation method should take about 4 to 6 hours. Hence, this must be done 29 times during this voyage, once for each instrument.

Given the raging ocean and wind, the previously unexplored ocean floor with a highly steep ridge, and the Antarctic Circumpolar Current, whose flow vectors at depth are unpredictable, it is hard to imagine any other place on land or ocean that would be more challenging than this. But thinking about the science that awaits us is more potent than anything and helps remove dark thoughts. Rocking with the beds with the occasional strong impact, in the small cabins under the oval ship's windows with the curtains drawn, we somehow drift off to sleep.

———

A mighty earthquake thundered and rumbled through this remote corner of the world underground, covered by a restless ocean, hidden from human senses and far enough away not to disturb a soul. Seismic waves created by the rupture of a vast underground rock mass began their long journey, tearing across the planet's surface and circling it for hours and days before dying out. The ground motion caused by these waves was recorded by seismographs tens of thousands of kilometers away. Geoscientists immediately noticed it, puzzled by such a monstrously large earthquake in an unusual location. After the initial analysis of the seismograms, it became clear: on 23 May 1989, an earthquake with a moment magnitude of 8.2 shook the Macquarie Ridge Complex in the Southern Ocean.

Australia's Macquarie Archipelago is surrounded by the tectonic complex of the Macquarie Ridge, along the boundary between the Australian and Pacific tectonic plates in the southwestern part of the Pacific Ocean and near the Antarctic plate on the southern side. Rising 410 meters above sea level, Macquarie Island is the only island on Earth made exclusively of oceanic crust and mantle rock (fig. 9.7). It is included in the UNESCO World Heritage Site primarily because of its unique geology and being home to colonies of birds, penguins, and sea lions. But why is this location, halfway between Tasmania and Antarctica, so special? Why did we choose this ghostly island and its inhospitable environment, hundreds of nautical miles away from civilization, out of all the beautiful places we could

FIGURE 9.7. The slopes of Macquarie Island covered in snow. We used the island's eastern (lee) side as a shelter from the strong westerly winds.

study on the planet? Captain Douglass called the island back in 1822 "the most wretched place of involuntary and slavish exilium that can possibly be conceived," where it is hard to imagine that any civilized soul would want to live. The reason for our choice is simple: in addition to its unique geological history, Macquarie hides, beneath the Earth's interior, the factory of the largest earthquakes on the planet unrelated to active subduction. Or maybe they are? Our expedition should answer that question.

Why do large earthquakes occur near a boundary where, once upon a time, tectonic plates moved away from each other, where fresh lava gushed up and pushed them aside? What physical conditions could have created the largest underwater earthquake of the 20th century in which two tectonic blocks moved sideways against each other? Are there indications that the blocks can also move vertically, and could future earthquakes cause a tsunami that would threaten the populations of Australia and New Zealand and perhaps all countries around the Indian and Pacific oceans?

What is the significance of the fifteen 20th-century earthquakes with moment magnitudes greater than 7.0, concentrated within a single tectonic plate, away from the edges? Is this perhaps a sign of the creation of a new subduction zone in this region? Indeed, understanding how plate margins evolve and subduction zones form remains one of the truly fundamental unanswered questions in the Earth sciences. The Macquarie Ridge Complex is central to that question because we know that the tectonic boundary between the Pacific and Australian plates, and nearby, the Antarctic, is one of the most dramatic tectonic boundaries in the world.

What exactly lies underground beneath Macquarie? This million-dollar question has slowly settled in me and won't budge. This time, it's not about the inner core—it's about something closer and much more tangible, something not so deep beneath the ocean floor.

A vast amount of the Antarctic Circumpolar Current has flowed over and around the ridge since the 1989 earthquake. Another strong earthquake, magnitude 8.1, struck the same area in 2004. Meanwhile, Australia has invested in purchasing 20 ocean floor seismographs, which have been waiting long enough to be used in an academic project like this—after all, that's why they were bought. I knew back in 2016 that I had to give it a go.

Writing project proposals is always more complicated than writing scientific articles or reports. Many things need to coincide:

a lot of patience, ups and downs, and small victories. A project like this one must win several tenders simultaneously: one for using the ship and its crew, another for using instruments, and yet another for the support of the research project, travel, and salaries of technicians and scientific associates. When you add restrictions due to the pandemic, a series of permits that had to be obtained from various authorities, health examinations, and quarantines, it becomes clear that this is not just a scientific endeavor.

Logistically, it was complicated to put everything together to get permits to work on such an inaccessible but protected part of the ocean near Macquarie Island. Considering that we overcame administrative obstacles during the COVID-19 epidemic, the success is all the more remarkable. It is a multidisciplinary, multimillion-dollar, and international project[13] for which we managed to get the support of the Australian Research Council, the leading state science institute (CSIRO),[14] that is, its leading blue-water research capacity MNF,[15] which assigned the RV *Investigator*, and the AAD[16] to place additional instruments on the island itself. We also added support from the British NERC.[17] Together with the seismographs, the material value of the project grew to more than AUS$10 million.

A few years after the first attempt at writing a project proposal, here we are on a modern research vessel with the best crew of scientists I could imagine and instruments ready to be

13. In addition to myself, the lead researchers on the project are Caroline Eakin (ANU), Mike Coffin (University of Tasmania), Nick Rawlinson (University of Cambridge) and Joann Stock (Caltech). For references, see the bibliography.

14. Commonwealth Scientific and Industrial Research Organisation.

15. The MNF is funded by the Australian government and operated by CSIRO on the nation's behalf.

16. Australian Antarctic Division.

17. Natural Environmental Research Council.

deployed. We came here with high expectations and hopes, armed with expertise and an arsenal of sophisticated scientific methods ripe for application to this problem. We will implore practically everything we have. To begin with, we will let 29 instruments listen to ground motions at preplanned locations on the ocean floor for about a year. They will record local, regional, and distant earthquakes, which we will use to study the subsurface beneath Macquarie. As a result, we rightly expect that the new data will advance our understanding of the propagation of waves in this part of the world and the physical reason for the earthquakes that occur.

It will be the first study to investigate the processes that generate the world's largest underwater earthquakes without being associated with active subduction. This could lead to understanding the subduction initiation process, the mechanism of earthquakes at converging plate edges, and better calculations of the potential for tsunami generation. It will serve scientists, legislators, and all people who live along the coast in maritime countries. This is a project with which we will try to answer the question of all questions in the field of plate tectonics: How did it all begin? How does subduction begin?

About three and a half days after leaving Hobart, we arrive northeast of Macquarie Island and begin sounding, describing lines parallel to the extension of the ridge. According to the plan, we should use soundings to create detailed maps of the ocean floor and determine the best locations on the ocean floor for our instruments. We prefer soft sediments so the seismographs can adhere firmly to the bottom with their weight. That is somewhat different from the choice we would make if we were somewhere on the continent, where we would look for solid rock instead of soft sediments because it gives a much better quality of ground motion signal. To get the best possible picture of the ocean floor, we must combine the results of broadband sonar and radar backscatter and acoustic mapping methods.

We sit in our computer lab deep beneath the deck and listen to the sound of sonar and waves crashing into the ship. If you've ever seen war movies, the submarines subgenre, you'll remember the sound of sonar that repeats periodically, with the tension in the air usually being cut with a knife. Being so deep beneath the deck in an enclosed space brings familiar feelings, but the constant motion and sound of waves remind us where we are.

Spending time during the night here, rocking amid furious fifties hundreds of kilometers away from civilization, leaves you with plenty of time to reflect on your life's calling as a scientist and a few other things, trust me. Peeking across a long desk, I admire the dedication and endurance of Thuany Patricia Costa de Lima and Thanh-Son Phạm, two of my graduate students affected heavily by seasickness but still working in these wee hours. Coming from Brazil and Vietnam to Australia to work with me on their doctoral degrees, little did they know where their academic path would bring them and how they would be united with me and the rest of the crew on this worthy endeavor that keeps pushing the limits of our endurance. For some time now, the three of us have turned a part of our subantarctic night shifts into brainstorming sessions, working on seismic waves of large earthquakes that traverse near the Earth's center on their journey through the interior.[18]

18. I am happy to report that since our voyage, we published several influential papers on the topic together. The same goes for the two other crew members, my graduate student Sheng Wang and postdoc at the time, Xiaolong Ma.

We want to get as close as possible to the slopes of the ocean ridge, but they are so steep and sharp that it is impossible to position the seismometers so that they adhere firmly to the bottom. Helped by the expertise of our marine geophysicists and co-investigators on the project, Mike Coffin, leading this voyage, and Joann Stock, supporting us remotely from California, we identify rare terraces of soft sediments on the steep slopes of the ridge. Because of the strong ocean currents, the instruments are carried laterally during the free sinking, so for them to successfully land on the sediments near the point we determined on the map, the soft sediment deposits must be at least several kilometers wide.

Strong wind and waves often force us to turn into the wind by placing the axis of the boat parallel to the direction of the waves, which is one of the strategies when you are on the open sea. The island is a natural wall against the westerly wind. The weather is miserable, so we often seek shelter behind the island's eastern side. Even with bad weather, the crew spots whales, sea lions, and penguins. My shift is until two hours past midnight, so unfortunately, I miss out on most of the sightings of the living world in the early morning. I would have to make an effort to be awake in the morning when the conditions for such observations are the best.

Despite constant winds of around 30 to 40 knots and occasional gusts of 60 knots, we found time yesterday to lower the first two seismographs, one of each of the two types we have prepared for this expedition. We lowered the first one on a wire to a depth of about 100 meters above the ocean floor before it was unhooked by an acoustic signal and allowed to sink freely. The accuracy of estimating the water column height between the instrument's acoustic release and the ocean floor is critical for this type of instrument descent. The estimation is done using sound waves, so the speed of sound in the water must be measured before each location.

The first instrument ended up at a depth of 4,525 meters. We had a more serious drama with another instrument, which, somewhere at a depth of 3,000 meters, broke free from the wire on which it was suspended and sank freely to the bottom to a depth of 5,500 meters, fortunately, without much lateral movement. After the seismographs ended up at the bottom, changing the ship's position, we determined their exact locations using the triangulation method. Each instrument installation took an average of 4 to 6 hours, as we predicted.

You can use just one seismograph to learn about an earthquake, as is the case with the current Mars mission, *InSight*,[19] or a whole group of sensors carefully designed into an array configuration, like the Warramunga array. Of course, each element of the array can be used separately, but combined, they can amplify weak signals from distant earthquakes like a giant antenna.[20] Motivated by the previous successes of the seismic arrays we had deployed in the interior of the Australian continent, my idea was to design a Macquarie Ridge Complex array that would extend in the form of three logarithmic spiral arms over kilometers with a center south of the island. This form enables the amplification of weak arriving signals, which are weakened by the presence of the ocean and atmospheric noise in this part of the world.

The northern part of the array is designed classically in consultation with my other two co-investigators on the project and seismologists, Caroline Eakin, who is on the voyage with me,

19. Acronym for Interior exploration using Seismic Investigations, Geodesy, and Heat Transport.

20. See chapter 8.

and Nick Rawlinson, who is supporting us remotely from Cambridge. Thus, the shape of this subarray is the letter X, which should allow us to create underground profiles across the ridge from the reflection and refraction of waves that reach our instruments steeply from the Earth's interior, that is, a series of layers with thicknesses and depths to their boundaries, velocities of wave propagation and density.

As we have seen already, the coverage of the Earth's interior by wave trajectories is limited due to the uneven distribution of seismographs and large earthquakes that generate waves large enough to be recorded on the other side of the world (fig. 9.1). Precisely because of that, the unique location of our carefully designed array should enable exciting research, including the "blue-planet seismology."[21] Besides, tracking storms and whales in the Southern Ocean are just some more modern research directions that come to mind.

———

Furious latitudes south of the 50th parallel live up to expectations and the reputation with which they earned their name. We have had to change our original plan several times so far. We have lost valuable time navigating to favorable positions upwind and leeward on the island's east side. The norm for wave amplitudes is a constant 4 to 6 meters, occasionally more than 10 meters, and for wind speeds between 50 and 80, with occasional gusts of over 100 kilometers per hour. The sound of the ocean water mass occasionally hitting our ship is not much different from an explosion. Sitting in the operations lab feels like sitting in a sound chamber. The movements are constant and

---

21. This kind of seismology is sometimes called environmental seismology.

violent, but most have already gotten their sea legs and learned to live with them. Or at least we have accepted that we cannot influence the weather.

Perhaps it is unnecessary to emphasize that the initially conceived shape of the logarithmic spiral array had to be changed on the fly as we progressed because of the extreme configuration of the ocean floor we saw. The terrain turned out to be too steep and dangerous, just as Mike Coffin, who once descended in the submarine *Shinkai 6500* down the steep slopes of the big island of Hawaii, had warned me. Indeed, if Macquarie Ridge were somewhere on the continent, it would be known as the most extreme mountain on Earth, that is, a steep wall 6 kilometers high and only about 40 kilometers wide.

Descending this ridge in a submarine would be like dipping down the steep roof of a cathedral. Let's compare Macquarie Ridge to the three largest mountain ranges on Earth by looking at the difference between their base and highest point. For example, for Mt. Everest in the Himalayas, the difference from the base to the highest point is about 3,700 meters, and its width is between 150 and 300 kilometers; for Denali in Alaska, part of the Cordilleras mountain range, it is 5,500 meters from the base to the top, but its width is 150 kilometers. Finally, for the peak of Aconcagua in the Andes, the difference is about 2,600 meters from the base to the summit, but it spreads over a width of 500 kilometers.

Mike, a man who has spent more than three years of his life cruising all the world's oceans on 35 previous voyages, warned me even before we started writing the project proposal that Macquarie Ridge is steeper than any other place on Earth. As such, it represents the most challenging terrain on Earth for setting up submarine seismographs because they require a flat bottom. But none of us expected that we would see evidence of the newly amassed ocean floor along the ridge's entire eastern side. With a length of about 100 kilometers and more, the whole mountainside had collapsed and ended up at the foot of this vast natural wall. A similar scene, although less dramatic, awaited us on the western side of the ridge. If one of the project's goals was initially to understand how much tsunami potential the earthquakes that occur here could have, now, the story has changed. Namely, such large underwater mass wasting and landslides can cause a catastrophic tsunami![22]

With the help of a transparent sheet of paper and a new bathymetric map, I moved and rotated our originally intended constellation of instruments in the form of a spiral to the east of the ridge, where the sonars recorded a deep but relatively flat surface of the ocean floor, about 5,500 meters deep. The shift had to be made so that all new locations coincided with the sections that we determined were sediments, which for many locations meant only a narrow terrace on the steep ridge slopes, a few hundred meters in diameter.

To my great satisfaction, we thus obtained new coordinates of the seismograph positions and, at the same time, kept the original shape of the network. However, landing the instrument in such a narrow area from about 5,000 meters above is a major logistical challenge. So far, we have recorded the lateral displacement of the sinking seismograph in all directions, up to several kilometers, which is unsurprising considering the eddy currents that occur behind the ridge when the ocean mass hits its high wall at a right angle. Considering that a strong circumpolar current carries the instruments laterally in unpredictable

---

22. We have not published this discovery yet, but we presented it at the American Geophysical Union conference in 2021. See Coffin et al. (2021) and Tkalčić et al. (2021a,b) in the bibliography.

directions, we got very narrow targets at depths of 5,000 to 5,500 meters for most of them.

———

When pressed by a deadline, you can only hope things go smoothly without a hitch. A 48-hour extreme storm was approaching, and we hoped we could get at least two more instruments set up before the waves got big enough to knock us entirely out of action. Element MRO21,[23] short for seismograph Macquarie Ridge Ocean element 21, did not differ in any way from other array elements. Strong winds of 40 knots and waves about 5 meters high meant that we had to calculate the motion vectors as best as possible for MRO21 to land successfully, preferably on the part of the surface of the ocean floor we had mapped out for ourselves. Careful calculations required strict selection and took into account many factors, including the strength of the circumpolar current.

We had previously given up on lowering the instruments to the bottom of the ocean by wire because we realized that the heavy acoustic transponder, a transmitter and receiver system weighing about 100 kg—customarily mounted on a wire above the ocean bottom seismograph—could damage the seismometer when lowered by a crane from the ship, touching the ocean surface together. We realized that because of the seismograph's buoyancy and the ocean's energetic motion, the transmitter could damage vital parts of the seismograph, which could also cause premature release of the ballast weights. On the one hand, lowering the seismograph with a wire and releasing it at

23. We subsequently installed five instruments on Macquarie Island, and their names begin with MRL, which stands for Macquarie Ridge land element.

a depth of about 50 to 100 meters above the bottom would save us precious time because we would not need to triangulate the exact location in that case. On the other hand, winding a wire weighing 500 to 1,000 kilograms on the ship's winch would take quite a long time. In the end, the decision was made to let the instruments sink without the wire, which meant that we had to triangulate the exact position of the ocean-bottom seismographs every time they hit bottom due to their lateral drift from the ship's location.

It was still quite early that afternoon. On the monitor hung from one of the partitions of the operations lab, the dramatic action of lowering the instrument from the deck into the ocean during extreme weather conditions could be seen taking place. The ship's crew and technicians moved on the deck like little dots in yellow raincoats and protective suits as the waves and wind washed them. The camera attached high above the deck was bathed in scattered water particles that the wind blew along the entire ship hull and above the observation deck on the top floor. Still, among the yellow dots, I could decipher the movements of Robert Pickle and Andrew Latimore, two of the familiar faces that came along with us from Canberra.

At one point, the sounds of approval and joy once again spread through the operations lab as we could see a yellow rectangular prism weighing more than 200 kilograms plunging into the restless surface of the ocean and slowly disappearing into its murky depths. We were relieved when we realized the instrument was heading to the bottom. At that location, the ocean depth was 4,785 meters, and with a sinking rate of about 35 meters per minute, we estimated that MRO21 would touch down in about 2 hours and 15 minutes.

The high-pitched sound coming from the direction of one of the sonar devices in the operations lab was reassuring. That meant we could communicate with the sinking instrument at a frequency of 11 kilohertz. I had already gotten used to these sounds and knew that when the distance between two sounds became approximately equal several times in a row, it meant that the instrument had landed. During that entire time, the instrument was carried by a strong current in the lateral direction on its way to the bottom, which meant that we were no longer directly above it. The last reading was from a distance of about 5,500 meters, and the instrument was still sinking. The difference between the two consecutive readings was gradually getting equal, and I was relieved as this could only mean that the MRO21 had bottomed out. Optimistically, I begin to check the route for the next seismograph so that the time to triangulate and move to the following location can be spent as efficiently as possible.

"It's coming back!" says the head of our electronic workshop, whose work expertise and dedication are irreplaceable. As if in a daze, I listen carefully for the gaps between the two sounds. Indeed, the gap is getting shorter and shorter, which can only mean one thing—our underwater seismograph had bounced off the bottom and started to return to the surface. I look through the oval window through which I often watched the waves on the horizon from our lab. The last ray of daylight is slowly fading. The sound of long waves breaks through the thick walls of the ship and mixes with the silence. I know we're in for a long night. The hunt for MRO21 has just begun.

About 40 minutes later, we send the command to release the weights on the submarine seismograph on its way to the surface. This command releases the weight ballasts and activates the sensor's flashing light. The weights had already been released in this case, so the command only activates the flashing light. We knew that the seismograph rises at a speed of about

50 meters per minute. That gave us a pretty good idea of when it will surface, but we still don't know exactly where. We continue determining the distance as long as the ship is 750 meters from the seismograph.

There is tension in the operations laboratory because the runaway seismograph MRO21 is getting closer to us. The summation of the ship's velocity and ocean current vectors and communication with the command bridge are becoming more frequent. We need to be as close as possible to the instrument when it appears on the surface and, at the same time, avoid proximity, because the captain needs enough room to maneuver in this stormy weather (fig. 9.8). If we get too close and pass it, it will take too long to turn this giant ship away and return to the instrument, which would have drifted westward due to the strong current. It is a game of cat and mouse.

The ship's acoustic receiver and transmitter system is just beneath the ocean surface and can communicate with the seismograph only when it is still in the water. When MRO21 stops answering calls, we will know that it has surfaced. Suddenly, there is a call from the command bridge, and in no time, there is confirmation that they have spotted its flashing light. I drop everything and hurriedly head down the main corridor, finding the door to the main stairs to the command bridge.

Hurrying up to the bridge, six floors above the laboratory, I remember my first days on RV *Investigator* and the agonizing ascent and descent of this same staircase on loose legs when the sea wolf in me had not yet awakened. I enter the open space of the command room in complete darkness. Fortunately, I know the layout of the objects and compartments; otherwise, I don't know how I would have coped. I walk around, tapping the smooth oval bulkhead, and now enter the vast open space of the command bridge, where the outlines of objects and human

FIGURE 9.8. A stormy night photo from the observation deck (the 7th level of RV *Investigator*, also known as the Monkey Island).

faces slowly emerge. The overall design of the bridge is impressive, with extended wings giving the master, first officer, and mates an unrestricted view of the operation. The console screens and lights can be seen arranged in a large semicircle. The master and the first officer greet me, pointing a finger at the flashing light on the open sea. Although the Moon is full, the clouds start covering it slowly, as if they are allied with us that night. Moonlight would make our work much more difficult.

The last time I read the position of the seismograph, I knew its distance was about 700 meters from us, but now, looking at its beam of light thrown by the waves and emitting in all directions, it is challenging for me to determine its actual distance. The westerly ocean current is around 2 knots, and the captain must react quickly to stay close to MRO21. At that moment, he lowers the bow propeller to improve maneuverability and get closer to the seismograph. Then he turns off all the propellers to avoid sinking the seismograph under the ship on his starboard side, from which we will try to capture it. My free estimate is that the approach takes about 25 minutes.

When the ship finally turns, the trajectory of the floating seismograph MRO21 coincides with its starboard side. On top of the seismograph is a complex hook custom-made in Canberra. Holding the long rods, which the deck crew has already pointed toward the wavy surface, is difficult, partly because of their weight and length, partly because of the stormy wind. Surveillance cameras show two familiar figures under yellow raincoats fighting the wind and, after several attempts, finally hooking the seismograph on their hips. They tie it to the ship's mobile crane and drag it to the stern. I note 22:40 as the time when the winch finally lifts it onto the deck. I witness a sense of relief and sudden enthusiasm. We all needed this little victory after so many lost days due to the stormy weather. The MRO21 rescue operation has given us an idea of what awaits us when we return one day to retrieve these instruments.

———

The weather is finally better; the wind has calmed down to around 20 knots, and with some luck, we will have the remaining six instruments down on the ocean floor and be on our way to Hobart by Friday night (fig. 9.9). From where we will lower the last seismograph to its one-year residence,[24] a 75-hour voyage to Tasmania awaits us. I write the last lines of reports for

24. With some mixed feelings, I have to add that some of our seismographs are still on the ocean floor while I am writing this, but we have high hopes that they will

FIGURE 9.9. Sailing in the "furious fifties."

Facebook and LinkedIn and try to save as much material and data as possible on disk. Images of the ship's cabin and narrow corridors, seismographs disappearing one by one from the deck, a stormy ocean, whales, albatross, aurora australis, snow-covered Macquarie Island (fig. 9.7), smiling faces of the crew (fig. 9.10), the ship's cooks, seasickness, my students and colleagues. Among these composite images, there are memorable moments captured by my shipmates that I missed while I was occupied with other tasks. And some scenes are engraved deeply in my memory. I save them in my memory drawers, knowing that the end of the voyage is quite near.

---

be recovered one day. However, this is beyond the scope of this story and, perhaps, inspiration for a sequel.

FIGURE 9.10. The research crew of the IN2020.V06 voyage after the voyage, from left to right: Andrew Latimore, Robert Pickle, Caroline Eakin, Thuany Costa de Lima, Yun Fann, Mike Coffin, Hrvoje Tkalčić, Rajesh Erigela, Xiaolong Ma, Sheng Wang, and Thanh-Son Phạm.

Seagulls and the Tasmanian coast have kept us company for a long time. The luggage is ready, and the cabin is cleaned. We slow down and sail into the port. Slowly, the faces gathered at the dock of the National Marine Facility with a welcoming banner are visible. The familiar image of hilly Hobart and the return of sounds that disappeared from our lives for a month. The children of a crew member wave at us from the waterfront. At that moment, it's hard not to think about those who are also impatiently waiting for us, and our eyes quickly fill with tears. It is difficult to describe the experience of returning after such an adventure. If the departure was emotional for many of us, the return is undoubtedly a different dimension.

A few weeks later, I wake up one warm night and stand in a room whose ceiling and walls are gently swaying. My legs are now straight, but the ground beneath me still dances. Where am I? Is my shift approaching yet? Are we still in the lee or sailing at full steam? I am reassured by the cry of a possum of the *Trichosurus vulpecula* species, which spends summer nights in a tree near our house wall, enjoying the taste of the sweet peel of a large ponderosa lemon. I peek into my son's room and see him sleeping tightly, hugged by a king penguin from a Hobart store, a friend for his growing-up days. IN2020.V06[25] was undoubtedly a voyage to remember, extreme geophysics in action and a feeling not easily forgotten. The preparations for the return voyage to collect the instruments are just beginning.

25. Here IN stands for RV *Investigator*, 2020 for the sailing year, and V06 for sail order number 6 in that year.

10

New worlds
and their explorers

"The universe is a pretty big place. If it's just us, it seems like an
awful waste of space."

—CARL SAGAN[1]

Buzz Aldrin and Neil Armstrong were ready to join the command module *Columbia*, which remained in orbit around the
Moon. They went the farthest of all, made that giant leap forward for humanity, for which lances were broken, walked on
the surface of the Moon, and collected samples. But they also
did something else, perhaps less known to the public—they

---

1. American astronomer and astrophysicist Carl Sagan was one of the most responsible for creating a new scientific field: astrobiology. His most important scientific
works are experimental studies on the subject of extraterrestrial life. The Pulitzer Prize
and the NASA Distinguished Service Award are among the numerous honors he received for writing and science communication. He was elected to the American Geophysical Union fellowship in 1972 and presided over its Planetology Section.

installed scientific instruments for "listening" to the lunar interior. The *Lunar Passive Seismic Experiment*, the first scientific package ever carried to the surface of another planet, included a seismograph that worked for three whole weeks.

While the two of them tried to get some sleep in the lunar module after the work was done and mentally prepare for takeoff—and I doubt that they could have fallen asleep because of the elevated adrenaline—the seismograph they set up recorded the first lunar quake and sent the data back home to Earth. To the overall satisfaction of all concerned, it was only the first of more than 12,000 moonquakes and other phenomena that would be recorded by the seismographs of the *Apollo 11, 12, 14, 15,* and *16* missions.

Each time Michael Collins found himself in the radio silence of the Moon's far side in the command module, the darkest thoughts would emerge from the depths of his conscience. What if he had to return alone back to Earth? How would he live with the burden of knowing that two of his colleagues got stuck down there, in the middle of a basaltic "sea,"[2] some 380,000 kilometers from their home in the United States? The seriousness of the situation was evidenced by Richard Nixon's prepared speech in case the only engine of the ascent stage of the lunar module *Eagle* had failed. Because if *Eagle* didn't take off from the Moon, he would have to explain to people that *Apollo 11* was—after all—a success. In the speech, among other things, it was written that fate wanted the two astronauts who went to the Moon as messengers of peace to stay there and that they would rest in peace as heroes of humanity.

---

2. "Seas" on the surface of the Moon are dark plains created by ancient impacts of large bodies and basalt outpouring. Earliest astronomers thought they were actual seas.

Luckily, the engine started, the *Eagle* soared high above the basalt-blue reflection of the Sea of Silence, and the three astronauts experienced in orbit the happiest rendezvous in history after more than 21 hours of eternity. Champagne was popped around the world, and Nixon's letter remained in his Oval Office desk drawer. Missions to the moon continued and took four more seismographs to four different locations in the following years. Aldrin and Armstrong later stated they thought they had about a 50% chance that the engine would start and run long enough to climb to the command module.

More advanced than the prototype seismograph from *Apollo 11*, the other four seismographs worked until 1977. In addition to seismographs, magnetometers were also crucial in discovering the lunar interior. We discovered a lot about it.[3] For example, we have shown that the tidal forces on the Moon, due to the Earth's gravitational field, are responsible for deep moonquakes in its interior. We discovered that its surface is crisscrossed by faults, just like Earth's, and that it has a core, with most likely a solid and liquid part, just as Earth does. We also discovered that its chemical composition is similar to that of the Earth's mantle and that it was most likely formed from the Earth's mantle after a smaller body, Theia, hit the protoplanet Earth.[4]

We started well but then stopped for a long time, as if we had had enough of the Moon, and turned all our attention to Venus, Mars, and other celestial bodies. Indeed, *Venera 13* and *Venera 14* successfully landed on the surface of our planetary sister

---

3. For example, see Wieczorek et al. (2006), Weber et al. (2011), and Garcia et al. (2011) in the bibliography.

4. See chapter 1.

Venus, but because of the hellish conditions prevailing, all the instruments survived for only a few hours, not only because of the temperature exceeding 400 degrees Celsius but also because of atmospheric pressure of 90 bar,[5] 90 times higher than that on the Earth's surface.

Under such conditions, the seismographs sent by the Soviets could concentrate only on recording noises on Venus. Rock fragments carried by the stormy wind in the thick Venusian atmosphere and the sound of expansion and decay of the probe material exposed to high temperature and pressure, unfortunately, made it completely impossible to interpret the recorded and transmitted seismograms in the short time that *Venera 13* and *Venera 14* managed to survive. However, from studying the ambient noise today, we can learn a lot about the internal structure of the planet, in some cases even more than from the seismic waves that propagate after an earthquake.

We reached Mars on the wings of the *Viking 1* and *Viking 2* probes, each carrying a seismograph. Unfortunately, these instruments were mounted on the top of the probes near one of the legs. They could record wind amplitudes and were designed in such a way that they could record only powerful quakes if they did occur. Until 1980, when contact with *Viking 2* was lost, apart from a few candidates that remained only possibilities, no feature had been recorded that could be said with certainty to have been a quake. It was a devastating result, but project leader Don Anderson, a geophysicist at Caltech, concluded from an analysis of the recorded noise that this did not mean that Mars was not seismically active. After all, that

5. The pressure of 1 bar equals 100,000 pascals, slightly less than 1 atmosphere, which is 101,325 pascals.

experiment served as a prototype for the InSight mission, which we will return to a little later.

———

Another rainy season is underway in the San Francisco Bay Area. While the crowd of 550 freshmen slowly pours into Wheeler Hall, I am hooded and not yet accustomed to the morning coffee that would surely have woken me up by now. I observe in front of the entrance a red leaf of *Ginkgo biloba* that has blended with the already decently soaked concrete of the university campus in Berkeley. I am thinking about the Red Planet, what students might ask me and surprise me with after today's lectures by professors Boering and de Pater about its atmosphere and future missions.

The Planets (Astro 12) course, in which I—with a first and last name that no American could pronounce at first—served as a lead teaching assistant, was among the top-rated among undergraduate students at Berkeley. I experienced it somewhat emotionally because even as an undergraduate student at the University of Zagreb, I was slightly torn between two possible directions for a doctorate in the United States. Geophysics, on the one hand, and planetary sciences, on the other, presented different options. Nevertheless, when I saw the offer from Berkeley for a doctorate on the Earth's inner core to work with one of the world's most eminent seismologists, Barbara Romanowicz,[6] that convinced me to come to Berkeley. During that semester as a teaching assistant, while I actively thought about Mars,

———

6. At that time, Professor Romanowicz and coauthors published a state-of-the-art paper in *Science* on inner core anisotropy. See Romanowicz et al. (1996) in the bibliography.

Venus, and other intrigues of the Solar System in the mornings, my full attention in the afternoons and evenings was directed toward the Earth's depths. In addition to the main project with the Earth's inner core theme, I also worked with Doug Dreger on the mysterious volcanic earthquakes of California's Long Valley Caldera.[7]

I enter the vast, largest auditorium on campus, already well filled by then, and sit in my usual place. I think about how the education system here is so liberal that sitting in this lecture, there are students who, in a few years after Berkeley, will be scattered around the wider world with degrees as lawyers, economists, psychologists, engineers, and scientists. They will have in common that in their first year, they could have taken courses about planets or dinosaurs, the popular courses Physics for Future Presidents or Mind, Language and Politics, an Asian language, or poetry by Robert Hass. There are top athletes among them, even Olympians, perhaps a little less daddy's sons than at one of the Ivy League universities on the East Coast or at the university across the bay, Stanford. There are also children of Nobel laureates who have Berkeley in their hearts and proudly wear their ring with the university's official seal that reads: *Fiat Lux*.[8]

A slightly thicker Helvetica with the following content stands out from the white canvas: *Mars: 11% of Earth's mass; 9 times more massive than the Moon; a considerable amount of mass resulted in slow cooling and prolonged volcanic activity (aiming at its*

---

7. It was common to have two projects as a PhD student. Unrest and a swarm of volcanic earthquakes in Long Valley Caldera coincided with the time I came to Berkeley. See Dreger et al. (2000) in the bibliography. In retrospect, these topics will define my scientific interests in the years after my PhD.

8. "Let there be light."

difference from the Moon). The Tharsis bulge—a surface of intense, long-lived volcanism. The lack of plate tectonics leads to very extensive cracking of the crust and lithosphere around the Tharsis area. The density of craters on the southern plateau is a sign of a much older structure than that on the northern volcanic plains.

The lecture ends by listing the goals of future missions on Mars: *1) search for signs of life from the past and present; 2) understanding climate changes on a longer time scale; 3) the search for natural resources and 4) the assessment of the potential of future exploration of Mars. 1996—Mars Pathfinder, Mars Global Surveyor; 1998—Mars Climate Orbiter (failed mission), Mars Polar Lander (missing in action), Nozomi (landed in 2004); 2001—Mars Surveyor: lander + rover; 2003—missions that return samples from Mars to Earth; 2005—1 kilogram of rock from Mars per a mission (2008).*

About 25 years later, when I write these lines, I am happy to say that we have records of many Martian quakes thanks to the seismometer SEIS, which was installed on Mars by the InSight probe. In addition to the results we already have, we still hope that the following months, as long as the current mission lasts,[9] will yield more quakes and new knowledge about the interior of Mars.

But what was our initial motivation based on? What made us believe that even after the relative failure of the *Viking* missions, we should reinvest the effort and substantial financial resources into another attempt? It is clear that questions like "Are we alone in the universe?" are the questions whose answers are more valuable than any investment and sending new probes to Mars. But are there mitigating circumstances that make it possible to believe that the conditions for recording

9. On 21 December 2022, NASA announced the end of the InSight mission.

ground motions on Mars should still be promising for another attempt to record quakes?

In an attempt to answer these and similar questions, let's jump back to the bottom of the ocean and to the question of why the coverage of the planet by seismographs is of essential importance.[10] The answer is simple for someone who deals with the Earth's interior—because, in this way, better coverage of the deep interior with the paths of seismic body waves from all directions can be achieved. We have already compared this situation with imaging the internal organs of the human body. Imagine that you are a doctor specializing in diagnostic or interventional radiology, and you need to make an image or a puncture, and your receiver works only partially. Or, for example, you suspect pathological signs, and you cannot do an abdomen ultrasound from all sides of the body.[11]

For a geophysicist, instruments on a specific part of the ocean floor mean an opportunity to listen to earthquakes or other phenomena, such as secret nuclear tests, the dynamics of glaciers, and underwater volcanoes. In recent times, we have learned, in addition to recording explosions, to seismologically listen to and analyze storms, whale migrations, the passage of meteors through the atmosphere, and even changes in ocean temperature.[12]

Since we are talking about the proliferation of instruments, this is an excellent opportunity to mention that today, in addition to classical seismographs, a few other types of instruments are used: telecommunication optical cables—whose total

---

10. We dealt more with this topic in chapters 8 and 9.

11. There was more talk about medical and seismic tomography in chapter 6.

12. The project in the subantarctic waters of the Southern Ocean that I wrote about in chapter 9 is directly or indirectly related to all these interests and goals.

length on the ocean floor reaches a million kilometers, and to which seismic sensors can be connected—and floating seismic sensors that are about three times cheaper than classic seismographs for working on the ocean floor. The latter instrument pool is called "sea mermaids" in our scientific community. After every major earthquake, mermaids send data via satellite in near real-time.[13]

Anyway, with various imaging methods, of which tomography is perhaps the best known, a snapshot of the Earth's interior is obtained by analyzing the recorded data. Once we know the Earth's structure, it is possible to theoretically predict the movement of waves through it and, thus, better understand the physical mechanism of earthquakes in that part of the world. Therefore, we hope that the data collected from the bottom of the Southern Ocean could help us figure out whether it is a new subduction zone that is just being born. Other goals include using seismograph records to study various natural phenomena, many of which, I suspect, we have yet to discover. In this sense, the proximity of Antarctica is undoubtedly promising, and surprises are possible.

Generally speaking, to study the characteristics of earthquakes, it is necessary first to make detailed maps of the underground of the place where they occur, similar to how we would study earthquakes in California, Chile, the Caribbean, Alaska, Japan, China, Indonesia, Iran, Turkey, Italy, or some other part of the world. That can only be achieved by setting up a network of seismographs that passively record the movement of the

13. My colleague, Frederik Simons from Princeton, the head of the Mermaids project, once said that classical seismographs for working on the ocean floor are Rolls-Royces, while floating mermaids are, in comparison, scooters that occasionally get involved in traffic!

ground due to the passage of waves from nearby and distant earthquakes and other phenomena. Lowering each instrument worth about AUS$200,000 apiece to the bottom of the ocean and that to a very narrowly defined favorable location in severe weather conditions on a ship whose use costs more than AUS$100,000 per day can be compared in terms of the complexity of the operation to the lowering of instruments to the surface of another planet. Surely, it is an endeavor two orders of magnitude cheaper, but the risk involved is enormous, just like the knowledge that can be returned to us.

And if you thought the problem of understanding Earth's interior beneath the Southern Ocean was complex, just imagine what it is about Mars with only a single instrument. Before the InSight mission, we didn't have a single recorded quake on Mars, and we weren't sure if there was volcanic activity beneath the surface. Moreover, no oceans, glaciers, or whales exist on Mars to record their movements with our sensitive sensors, so we can boast about some exciting data in case no earthquakes are recorded. In fact, of all the phenomena that we could call "terrestrial," we knew we could record wind and possibly a meteoroid impact. So, how do you write a competitive project proposal for setting up a seismograph on Mars?

To answer that question, imagine what we could discover about a planet whose interior we still know practically nothing about! This knowledge, or lack of knowledge about Mars, is the driving force behind everything. Why is it even interesting to us? What are the motives that inspire many scientists around the world to devote themselves to the interior of Mars and other planets? The answer may surprise some with its simplicity: it is human nature to investigate what we know nothing about and whose secrets are worth revealing from the perspective of someone interested in the evolution of the Solar System and the possibility of inhabiting Mars in the future. The project proposal and the approved mission are the results of a 30-year effort by colleagues from NASA, JPL,[14] geophysicists, and planetary scientists worldwide. But academic reasons aside, the US$830 million cost of the InSight mission is still a lot of money that must be justified.

Imagine you were the project manager who had to convince NASA that this project should interest the broad scientific community and the public. Looking from the perspective of a scientist, but also from the perspective of a reviewer, it is always more challenging to write a project proposal in which you need to "sell" a cake that you haven't even baked yet than to report the results of an experiment that you have already conducted in a scientific article. There's always a fine line between eternal fame and having reviewers flay you alive or break your heart. And so, as Bruce Banerdt, a middle-aged Mars Exploration Program scientist at JPL in California, pressed the *submit* button, he wondered if he had forgotten to take advantage of an important detail that could ultimately cost him the success of the InSight mission project proposal. Something he and his team may not have included in the eight or so key questions about the structure and internal dynamics of Mars that InSight would have to answer.

The early years of the Solar System, the first few tens of millions, are recorded in the structural shells that Mars hides deep in its interior, and they are, it can be assumed, relatively preserved given a less turbulent geological past than that of Earth, which erased a large part of the traces. That is why research and dissemination of knowledge about the formation and evolution of the planets of the Solar System are the project's primary

---

14. Jet Propulsion Laboratory.

goals. Because InSight didn't go to Mars to explore its surface, but what hides deep beneath it, a kind of time capsule buried in the interior of a planet far from our blue dot, an impossibly distant 225 million kilometers. Considering that the project proposal was a success and that a total of more than US$800 million was allocated for it, Bruce Banerdt likes to say today that the cost of each of the eight critical questions from the project proposal is about a hundred million dollars.

It is difficult to dwell too long here on each of those eight essential questions that "sold" the project proposal to the broad community of planetary scientists, but something more could be said about one of them: the question of the core of Mars. Namely, from the measurement of gravity and other parameters of Mars from a distance, the precession and nutation of the rotation axis of Mars can be determined, just as is the case for the Earth.[15]

From these painstakingly collected astronomical and geodetic data, it is possible to make sketches or models of the interior of Mars with various cores, changing their size and density so that the models finally fit the observed data. However, while for the Earth—thanks to seismology—we can say with a precision of a few kilometers at what depth its core is, for such a small planet as Mars, in the absence of seismographs on its surface and considering only gravity measurements, the depth at which the core begins is relatively diffuse and indefinite in an interval of some 600 kilometers. In addition, it is impossible without seismological observations to say whether, in the very

15. Precession and nutation together define the path described by the rotation axis of Mars due to the gravitational interaction of the Sun, its satellites, and other planets with Mars. See more details in the glossary of relevant terms.

center, this core is perhaps in a solid aggregate state or is entirely liquid.

Mars is fascinating to us because, because of its smaller size and somewhat less turbulent geological history than the one experienced by Earth, it is an excellent planet for understanding the process of planetary differentiation, which took place very early in the history of the Solar System, and the traces this process left behind. Although we do not have rock samples from Mars brought back to Earth by astronauts as is the case for lunar rocks, indirect age measurements can be made by observing the number and patterns of craters on its surface and age can be determined by measuring the ratio of various isotopes for meteorites that have fallen to Earth.[16] But the problem is that it is impossible to determine which part of the surface of Mars they were broken from. Besides, they also became contaminated with other chemical elements by passing through the Earth's atmosphere.

To investigate the possibility of life on Mars and the composition of surface rocks, NASA and ESA jointly operate the *Perseverance* mission as a logical continuation of the *Curiosity* mission. If the previously mentioned US$830 million seems a lot to someone, Perseverance costs about US$2.73 billion. In addition to various scientific measuring instruments, it also took the *Ingenuity* helicopter to Mars. By drilling, the rover collected its first sample of the Martian rock column in September 2021. According to the plan, these instruments will collect more such samples from different parts of Mars. They will be launched into Mars orbit, where they will be picked up by a probe that will bring them to Earth. The mission has several

16. In addition, rock samples are stored onboard the Perseverance rover. NASA plans to transport them to Earth in the future.

stages, each crucial for the successful return of the samples home. It is no exaggeration to say that the InSight and Perseverance missions are collecting time capsules buried for millions of years under the rusty surface of Mars. Do they hide signs of life, the first proof that people like Carl Sagan are correct and that we are not alone in the universe?

———

Should we expect quakes on Mars at all, to what extent, and how do they occur? Quakes on Earth are caused primarily by tectonic forces due to plate tectonics, and without active tectonics, it is thought that Mars is a relatively calmer planet than Earth. Although Mars does not have plate tectonics, its surface does not hide the scars of a turbulent and not-so-long-ago geological past. For example, in geological terms, Olympus Mons, the highest elevation in the Solar System, was formed not so long ago.[17] This dormant stratovolcano rises as much as 22 kilometers above the surface of Mars. Its size is definitely a sign that once in that place, the vertical rise of basalt magma from the mantle "fed" the stratovolcano while the crust of Mars was fixed in one place. If the tectonic plates happened to move laterally on the planet's surface, as is the case on Earth, then the "scars" on the surface would look more like a volcanic mountain chain similar to Hawaii than a solitary volcano.

Since there is no plate tectonics on Mars, what can cause rocks to shear or collapse? One possibility is a change in the gravitational field by the tidal forces from the two moons of Mars, Phobos and Deimos. Because of Phobos, tidal forces deform Mars by about 1 centimeter with each oscillation. In addition,

17. According to the crater count method, its age is just above 100 million years.

the crust of Mars is exposed to diurnal temperature variations, which together cause a changing pressure field in it. These oscillations can contribute to the cracking and faulting of rocks and the occurrence of quakes similar to moonquakes, also possibly on Jupiter's satellites. In addition, the impact of a meteorite on the surface causes a considerable excitation of elastic waves, as recorded on the Moon. For those unfamiliar with the meteorite fall phenomenon, Mars suffers a relatively large number of these phenomena due to its thin atmosphere compared with Earth's. Practically every pebble that would burn up in the Earth's atmosphere reaches the surface of Mars.

Of course, it is right to ask how much a single seismograph of the InSight mission can help in finding answers to such important questions, especially when we know that we have tens of thousands of seismographs on Earth, and we are still in the dark when it comes to an understanding of its deep interior. For example, the triangulation method is used to determine the location of an earthquake, for which we need at least three instruments to determine the earthquake's epicenter with certain assumptions. How can we calculate such a seemingly trivial parameter as the epicenter with only one sensor? The answer to this question is hidden in the data we can record. However, there are also several mitigating circumstances that, fortunately, did not conspire against us.

First, the *Viking* seismographs were located on the probe's hull, high above the Martian surface, resulting in a catastrophically poor record. One of the sensors that operated recorded mostly atmospheric disturbances and oscillations in pressure and temperature, prompting scientists and engineers to design InSight's seismometer to be as close to the surface of Mars as possible, just like that ginkgo leaf that made me think of the Red Planet. At the same time, its design insulated it from Mars's atmosphere.

Mars has a relatively thinner atmosphere than ours, consisting primarily of carbon dioxide, but it also has circulation from diurnal to global. Mars is ravaged by sandstorms and numerous dust devils, and what a seismometer can most easily record are just such oscillations in atmospheric pressure. That's why the engineers made a cover that would isolate the sensor from atmospheric influences this time around. The InSight mission was delayed for more than two years because one of the three layers of thermal covers specially made for this mission had a microscopic hole that let in the "air" of Mars. It was so small that you would have to wait about 50 years to see a change of about 7,000 pascals, about 30 times less than the pressure of your car's tire.[18] However, even such a small hole was extremely important from the team's point of view and had to be "patched" if the seismometer would do its job as perfectly as possible. Given so much caution and interventions of this type, we certainly have better technology now than we did at the time of the *Viking* missions.

Second, Mars is much smaller than the Earth, which means that the energy of elastic waves if a quake occurs, or the waves that spread across the surface of Mars and through its interior after a meteoroid impact, lose energy much slower than on Earth. That means that even a relatively smaller and shallower quake, say a moment magnitude of 4.0, would generate enough energy for the surface waves to circle the planet several times, as with major earthquakes on Earth.

Imagine having only one seismometer on the planet's surface (point A) and a quake at any distance from it (point B). It is then possible to draw a circle on the surface around the planet's center so that points A and B sit on that circle. We will then have

18. A typical tire pressure is about 220,000 pascals, slightly over 2 atmospheres.

a shorter arc connecting points A and B[19] and a longer one, which we would draw if we looked in the opposite direction. Surface waves travel in all directions from the epicenter of the earthquake or the ground zero of the meteorite impact and also along those two arcs. We can predict their absolute and relative times quite accurately if we know the speed of propagation of the waves in the surface rocks of Mars. That is possible because we have already studied their patterns. Precisely, from the combination of predicted travel times and the behavior of spatial and surface waves along shorter and longer arcs, it would be possible to determine the location of an earthquake from a single seismogram without the mentioned triangulation method.

Third, apart from a relatively thin atmosphere, Mars has no oceans. On Earth, due to the interaction between the atmosphere, the oceans, and the solid part of the planet, there is a continuous noise consisting of a broad spectrum of frequencies, what we have already defined as microseismic noise or disturbance. Although the Martian atmosphere still generates noise, it can be expected to be much smaller than Earth's. Indeed, the first records of ground motion from the surface of Mars revealed that microseismic noise is about three orders of magnitude lower than on Earth, especially at night. That means Mars is a relatively quiet place, and it should, therefore, be easier to "listen" to signals from its interior. In addition, it can be statistically estimated that the instrument could record between 20 and 30 meteoroid impacts in two years of operation, the time interval for which financial support was anticipated.

It should be clarified that InSight has two seismometers under one cover. One is a short period, which means that it is

---

19. In a similar way that an airplane flies from point A to point B by the shortest route.

sensitive to movements of relatively shorter periods (higher frequencies), and the other is a broadband instrument sensitive to a very wide interval of periods, that is, frequencies. In addition, each of them has three channels or components. Three mutually perpendicular components are necessary to describe the three-dimensional motion of ground particles in space, and this is the instrument's sensitivity to an extensive range of motion frequencies. On the one hand, there are very long periods of about 100 seconds, which correspond to the free oscillations of the planet after large earthquakes.[20] On the other hand, there are very short periods or high frequencies, in this case, up to 20 hertz, which means that the sensors can simultaneously record ground oscillations caused by dust devils, small quakes, and meteoroid impacts.

As we have seen, due to the turbulent geological history, it is logical to expect that the Martian crust is cracked and is cut by fault planes comparable to the faults on Earth. One such fault system is Cerberus Fossae, some thousand kilometers away from the InSight lander. Hence, many have wondered why the location where the InSight probe landed is relatively far from potentially more interesting geological localities on Mars, such as Cerberus Fossae.

However, from the point of view of seismic wave propagation, it is more relevant to move away from interesting locations where multiple earthquakes may be happening and place the instrument at some decent distance from them because that way, the body waves from the quakes to the sensor will travel deeper through the Martian interior. We have seen how this is also true for Australia, a continent that is stable and, in

20. Due to its internal structure and size, the planet oscillates at specific frequencies like a bell or a stringed instrument.

principle, seismically inactive, but due to major earthquakes in Indonesia, in the area of Tonga and Fiji, New Zealand, the Kermadec Islands, and other places in the surroundings (see figures 6.1 and 9.1), it is a phenomenal natural laboratory for the study of the subterranean architecture of the Earth directly beneath the continent.[21]

So, InSight landed on the plain of Elysium Planitia, an area of Mars that may not stand out for interesting details on the surface but was carefully chosen for the safe descent of the probe as a place from which we will peer into the interior of Mars. There should have been no mistakes during the descent because the most important device of this mission, its seismograph, needs to be placed on the ground in a flat place to be efficiently pressed against the rusty solid body of Mars.

———

Marsquake, marsquake! Not so far from us, looking from a space perspective, on our neighboring planet, similar but still somewhat different phenomena from those on Earth are taking place. In 2019, we could read this breaking news in newspapers and published articles after the first digital data of ground motion recorded by the SEIS seismograph became available to the team of scientists who had access to the data before they were released to the rest of the scientific community.[22] It could be said that the very news that there is a marsquake on Mars justified the investment in the InSight mission because, as we have seen before, marsquakes were never documented. More precisely, the *Viking 1* seismograph did not even activate due to

21. See more about this topic in chapter 8.
22. See some relevant publications in the bibliography.

technical problems. The *Viking 2* seismograph, placed on top of the probe and active for more than 500 sols,[23] unfortunately did not record any event that could be safely said to have been a Martian earthquake, except for the strong wind.

Now, let's recall what we learned about ground motion records or seismograms and imagine that we are among the members of the team who receive fresh data from Mars daily. The horizontal axis of the seismogram shows time, and the vertical axis shows the acceleration of ground particles in meters per second squared. Let's also recall the spectrogram. With it, on the vertical axis are the recorded frequencies in hertz, and different colors indicate the relative strength of the recorded frequencies. The places in the spectrogram where the signal is particularly strong are marked in red, while dark blue indicates the signal is weak or absent.

On the spectrogram of the first marsquake, which was released to the public, we immediately saw that the sensor was seldom completely still. Due to wind and atmospheric stimuli, as well as the instrument's configuration, some frequencies are more excited than others and are constantly present. They are seen as yellow-green horizontal bands, for example, at around 4 and 6 hertz, which, in simple language, means that the oscillations recorded by the sensor are at frequencies of four and six times per second. The signal sinks into the blue and disappears at around 8 hertz and higher frequencies. However, a sudden change occurs at one point on the 128th sol after the descent of the InSight probe. The ground particles begin to "go crazy" at

---

23. Sol is one solar day on Mars, i.e., the time it takes for Mars to rotate once around its axis, an average length of 24 hours, 39 minutes, and 35 seconds. This is similar to Earth's day; however, one solar year on Mars (the time it takes for Mars to orbit the Sun) lasts 668 sols.

almost all higher frequencies, especially between 5 and 8 hertz, for nearly 10 minutes. The lower frequencies quiet down first, while the higher frequencies quiet down the slowest, thus forming a kind of "tail" in the spectrogram, giving it the shape of a fish.

At the same time, we could also see a seismogram, and in this case, the acceleration of ground particles as a function of time. We should remind ourselves of elementary school physics and that on the Earth's surface, the gravitational acceleration of objects due to the Earth's mass is about 9.81 meters per second squared, which is sometimes also called 1 g. The acceleration of the Martian earthquake on the vertical axis showed us that SEIS recorded amplitudes about ten times less than that. Large earthquakes on Earth cause ground motion accelerations that approach values of 1 g or, although rarely, are even higher. But let's not forget that Mars has a smaller mass than Earth, and the gravitational acceleration on its surface is only about 3.7 meters per second squared!

However, the most important thing to note here is that from the wave packet in the seismogram, which appears at the same time as the energy that we identified from the spectrogram as a marsquake, neither P nor S waves nor surface waves can be distinguished by the eye. Because of this, it is impossible to determine the distance to the marsquake or its exact magnitude. The absence of characteristic waves may be explained by the fact that this marsquake is relatively small in magnitude, and it is also possible that it was not a classic tectonic quake.

The conclusion is that this Martian quake looks more like a lunar quake than an earthquake! Namely, in lunar seismograms, it is also the case that the amplitudes of motion decrease very slowly, 10 or more minutes after the arrival of the first waves, which is very different from terrestrial quakes. In other words, the "earthquake tail" is much shorter than the "marsquake tail"!

The reason for these fundamental differences, apart from the physical mechanism of the quake itself, should be found in the material through which the waves pass. It is possible that the architecture under the surface of Mars is very heterogeneous and that, due to the scattering of waves from these small heterogeneities, their amplitudes weaken much more slowly. Some also believe that the possible presence of water in the mantle of the Earth can result in a substantial weakening of seismic waves, which is not the case for the Moon and, apparently, not for Mars.

Moreover, when subsequent data analysis and the results were published in the following months, it became clearer that most marsquakes with long tails have a source in the crust, probably at several common locations that, unfortunately, cannot be unambiguously determined for this type of quake. Namely, it turned out that their waves spread to SEIS exclusively through the crust as a kind of conductor, but their energy is so scattered that it is impossible to determine the exact direction from which they approach SEIS.

They got the name "high-frequency marsquakes" because their seismograms look like a series of lines that someone with a convulsive hand crossed in a vertical direction on a horizontal line of paper. They have long tails and are, in nature, something between terrestrial and lunar quakes. There are hundreds of them. There is also a subtype of local, very high-frequency quakes, which have been shown to have a thermal origin. Namely, due to the contraction and expansion of rocks during sudden temperature changes, for example, at sunrise and sunset, the seismograph records very high frequencies of these phenomena.

At the very beginning, in addition to the spectrogram and seismogram of the marsquake from sol 128, several other possible candidates for marsquakes were published. Particularly interesting were the signals of the monochromatic type, which means that they excited only specific frequencies of ground motion. At the time, this seemed improbable because no trivial physical mechanism could explain such a phenomenon. The million-dollar question in those days was, what does a meteoroid impact look like, and is it a characteristic structure of Mars that acts as a guide for only specific wave frequencies? Or maybe it is about periodic "breathing" and moving magma through a volcanic pipeline deep below the surface of Mars? After all, whether it was a familiar phenomenon or something else had yet to be investigated.

About a month and a half later, on the 173rd sol, we received a prototype for another type of marsquake: these are the so-called low-frequency marsquakes, whose records look as if, instead of a series of lines going up and down, you crossed a horizontal line of paper with wavy lines. This second type of marsquake is extremely interesting to us because they look more similar to earthquakes, but we still don't know much about them. Were they caused by tectonic processes on Mars, or perhaps by the movement of magma in its interior? They are deeper than the first type of marsquakes, and their foci are very likely in the mantle or at the bottom of the crust. So far, only a few of them have been observed with good enough quality to recognize their P and S waves, study how ground particles move in space, and determine their location. Their predominant location is in Cerberus Fossae, a system of parallel cracks in the crust.

This fascinating system of parallel faults and fissures in the crust that extends over 1,000 kilometers is most likely caused by stresses due to the volcanic activity of Tharsis to the east and Elysium Mons to the northwest. Lava deposits estimated to be only 50,000 years old testify to relatively recent volcanic activity that could continue even today. Interestingly, these low-frequency marsquakes were observed and documented in the first reports only during the night on Mars, which is seismically quieter than the day. Conventional detection methods work so that they can only identify earthquakes that are more pronounced than ambient noise, and these conditions only exist on Mars for a few hours during the night. That is why it is unsurprising that all the low-frequency marsquakes recorded in the first two years by the team that worked on them were nocturnal. The first research papers concluded that it was most likely tectonic quakes in the Cerberus Fossae system, although volcanic causes could not be excluded.

In the meantime, the catalog of marsquakes has become more and more extensive, and it now contains hundreds of earthquakes of the first type and several tens of earthquakes of the second type, but only a few marsquakes of good quality—with moment magnitudes around 3.6, where P and S waves can be read. The best two provided essential knowledge about the thickness of the Martian crust and the profile of the mantle and confirmed the core's existence.[24]

Then 2022 arrived, and with it, our paper about marsquakes, the research we started just before the COVID-19 pandemic because, by that time, the waveform data were released to the wide scientific community.[25] We reported 47 hitherto undetected long-period events, most likely marsquakes, which testify to the fact that the interior of Mars is not calm but, like the

24. See Khan et al. (2021), Knapmeyer-Endrun et al. (2021), and Stähler et al. (2021) in the bibliography.

25. The authors are my colleague, Professor Weijia Sun from the Laboratory of Earth and Planetary Physics of the Chinese Academy of Sciences and myself. See the publication Sun and Tkalčić (2022) in the bibliography.

Earth's mantle, in a state of slow convection. What we did differs from the methods of a team of analysts who carefully studied seismograms and recorded each detection in a catalog. We approached this problem using two unconventional detection methods: let's call the first of them Benford's law method[26] and the second the matched-filter method.

According to Benford's law, in a data set, most numbers start with the digit 1, and the fewest begin with the digit 9. In fact, the distribution of the first digits behaves according to a specific law regardless of the number system used. Suppose you notice the first digits do not act according to this law. In that case, you may have detected an irregularity or even a deliberate entry of incorrect data into the database, which is used to verify bank accounts, the regularity of counting ballots, and similar applications. Or, if you apply that law to nature or the universe, it is possible to notice an earthquake or other phenomenon just based on the first digits of the digital data without studying the seismogram itself.[27]

But Benford's law does not work well here because the Martian ground motion data set does not extend over enough orders of magnitude; in other words, the dynamic range is not large enough, which prevents the detection of deviations from

26. According to American engineer and physicist Frank Benford.

27. My colleagues Malcolm Sambridge, also from ANU, Andy Jackson from ETH, and I published a paper in 2010 in which we detected an earthquake for the first time without seismogram visualization using only Benford's law of the first digits. In that paper, we confirmed that Benford's law is valid for large geophysical, astronomical, and other data sets. The magnetic field of the Earth, the gravity model of the Earth, the distribution of the depth of earthquakes, the rotational frequencies of pulsars, the masses of the discovered planets outside the Solar System, the number of cases of infectious diseases, and the emissions of greenhouse gases by the countries of the world are just some of the data sets that we have investigated. See Sambridge et al. (2010) in the bibliography.

Benford's law. On the other hand, the matched-filter technique worked great!

The principle is to cut out about 20 seconds of seismogram of a known marsquake like a cookie cutter or fingerprint, and then compare it with all the other data you have moving forward in time, second by second, minute by minute, hour by hour, month by month, until you have crossed the entire year for which you have data. More precisely, the pattern is centered on S waves, which are more pronounced than P waves. Suppose the similarity in the waveforms of that pattern and the samples you are examining at some point in time is significant enough. In that case, you have likely detected a replica of the marsquake that occurred at the same location and with the same physical cause. This way, we detected 47 new low-frequency Martian events that were likely quakes in the Cerberus Fossae area.

The number of 47 newly discovered events is greater than all marsquakes of this type detected before our paper was submitted. What is the importance of our discovery? First, if these events are marsquakes, Mars is a much more geologically active planet than we thought. Recurring marsquakes most likely mean that they are of volcanic origin, either due to the movement of magma in the Martian mantle or, more likely, due to periodic processes of gas release from the magma, which from time to time increase the pressure and cause cracks to open and close, like some of the extinct volcanoes on Earth. Also, we found that those events occur at all times of the Martian day and night, meaning we can rule out the tidal forces of Phobos and Deimos as causes. It also means that we can rule out thermal causes for this type of marsquake, which would recur periodically due to a sudden temperature change at the transition from day to night and vice versa.

If the Martian mantle is active, as our discovery suggests, this sheds light on the absence of a magnetic field on Mars. Namely, if there is slow convection in its mantle, as suggested by volcanic marsquakes, then convection in its liquid core is also more likely due to the efficient flow of heat through the interior of Mars. However, since the dynamo was once active and now is not, this may mean that the liquid core layer in which convection happens is relatively thin, so it cannot generate and maintain a significant magnetic field.

Alternatively, if there are a lot of lighter chemical elements in the liquid iron of the Martian core, for example, hydrogen and sulfur—and there is evidence for this—the liquid core did not mix well but segregated iron and sulfur in one layer and iron and hydrogen in another layer. That immiscibility, in turn, had disastrous consequences for Mars because it stopped the convection, thus the dynamo, and the protective layer by which Mars was shielded from cosmic radiation.[28] It remains to be seen if we can identify two layers in the core of Mars from new data to test this hypothesis. By the way, that "unfortunate" scenario did not occur on Earth because of a different ratio of lighter elements that reached the core during differentiation and because of different temperatures and pressures at the depths of the outer core.

Be that as it may, Mars, the son of Juno and Jupiter, blushes in the night sky; for centuries, it attracts and hypnotizes with its existence and makes one think about the mysterious expressions on its face. A complete understanding of its surface and interior, tectonic and volcanic activity, magnetic field, and the moment in history when it died out are crucial for future missions and for establishing human colonies on Mars. We can

28. See Yokoo et al. (2022) in the bibliography.

only hope that some of the forensic methods and algorithms we have developed to detect earthquakes and study the interior of planets, including their sediments and ice sheets, will be useful in future missions. From that point of view, our research and results are a small step forward, which is gratifying, but we still have a lot of work to do!

If someone had predicted to me as a boy who still remembers the descent of the *Viking* probes that one day, as a scientist, I would participate in analyzing marsquakes and also measure the Martian core's size using an innovative method,[29] I would have told them that they were dreaming. But humans have learned that we can and should dream big and wonderful things. We expect to send astronauts to the Moon again and instigate the journey of the first astronauts to Mars.[30] Missions to distant planets and smaller bodies of the Solar System like icy moons await us. As I write this, the science teams of the InSight and Perseverance missions are busy, and it won't be long before you see news of a discovery somewhere. We also need a bit of luck to record more marsquakes and other phenomena so that their physical causes can be studied and clarified.

———

At the time when, 28,000 years ago, our ancestors first made a figurine of a human body instead of a food container out of clay, a ray of light began its long journey from the galactic center toward us. It is an unusually long time from a human perspective,

---

29. See Wang and Tkalčić (2022) in the bibliography.
30. I didn't include the discoveries made after 2022 in this chapter. There is also the news that Australia will deploy a seismometer on the far side of the Moon in 2026!

but only the blink of an eye in the long life of the universe. As light traveled, we conquered the planet and created the first civilizations. We made a furnace, a wheel, a plow, a calendar, a catapult, a telescope, a steam engine, and a seismograph. Wars have been destroying us, natural disasters, pandemics, and famine, and human foolishness has always thrown us back. Despite everything, we managed to take more steps forward than back, create, and be productive; numerous scientific and artistic works were created. We studied and learned a lot.

The exploration of our home planet, the planetary system, and the universe as a whole define us as a human species. All the knowledge we have, all the methods we apply to research the interiors, surfaces, and atmospheres of other planets, all the models of their interiors that we have made, and also the seismological reconstruction of the deep interiors of the Sun and stars—helioseismology and asteroseismology,[31] we have learned and perfected on the example of our planet, of our natural laboratory. Curiosity drives us and continuously pushes us to take risks, to go millions of kilometers away from our laboratory in an effort to explore the secrets of other planets and bodies of the Solar System.

We have always had that deep urge to question, explore, push the boundaries of what is possible, and go one step further. That was true whether it was some member of *Homo erectus* who from an early age wanted to explore the mountain looming far on the horizon from their cave, or a member of *Homo sapiens* who conquered the Earth and now aspires to step onto Mars

---

31. Only P waves (also known as longitudinal or compressional waves) propagate through gas and plasma (electrically charged gas from which stars are built due to high temperatures), unlike the solid Earth, through whose interior both longitudinal and shear waves can propagate.

and soar even further, into the vastness of the giant planets and frozen frontiers of the Solar System. Because something is there, that's why we must explore it. Because the inner core is at the center of our planet, we very much want to understand and explore it, like a planet within a planet. Because a storm is raging in Jupiter's atmosphere, we want to study and understand it. Because rings surround Saturn, we want to take a closer look at them. Because the ocean may lie beneath Triton's[32] ice sheet, that's precisely why we want to find it. Because life may be thriving in the saltwater ocean beneath Enceladus's icy crust, we're racing to get there first.[33]

Unquestionably, the urge to explore other worlds continues the realization of the human species' potential. In that process, we realize ourselves, but we are also practical: we make progress in technology, create new industries, raise new generations of scientists and engineers, and nurture international relations.

And you know what? That ray of light has finally reached us after 28,000 years of travel, and we are still learning and can rejoice in the discovery like small children of a new toy. If it could be said that we have entered the era of mapping the surfaces of our own and other planets, symbolically speaking, we are still in the age of great discoveries about their deep interiors, origin and evolution, current dynamics, and cosmic destiny. We are still pioneers and conquerors of that part of the unexplored universe. We do not yet know where the research that has taken off, especially in this decade, will lead us. We don't yet exactly know which planet we will point our finger at and call ideal for

32. The largest of Neptune's 13 satellites. In Greek mythology, the god of the sea.
33. One of Saturn's satellites. In addition to NASA, private investors are also interested in missions to Enceladus. In Greek mythology, the son of Gaia and Uranus.

settlement. We don't know which planets are our alternative options. Meanwhile, the dream of covering the Earth with seismographs and other scientific instruments is getting closer to reality.

So, what do we do next? This question lights up my face every morning, as it has ever since I began to feel my calling deeply.

GLOSSARY OF RELEVANT TERMS

anisotropy—property of a material to depend on direction. In the context of solid Earth geophysics and seismology, seismic waves propagating through an anisotropic medium have different travel speeds for different propagation directions.

antipodes—points on the diametrically opposite sides of a circle or sphere. In geography and global seismology, a diametrically opposite point on the Earth's surface.

asteroids—celestial bodies of the Solar System larger than 1 meter that orbit the Sun mainly in paths between the orbits of Mars and Jupiter. They are made of metal or rock and have no atmosphere (unlike comets).

asteroseismology—a branch of astrophysics that deals with studying the interior of stars using seismological methods and principles.

astronomy—a natural science that deals with the study of everything above the Earth's atmosphere, including the celestial bodies, the space between them, and the entire physical universe and the phenomena in it.

attenuation—reduction of signal strength, e.g., the amplitude of seismic waves during transit due to absorption caused by the microscopic structure of the material, boundary surfaces, and crystal dislocations, and due to scattering caused by the presence of heterogeneity in the Earth's interior.

aurora (aurora borealis)—an optical and sound phenomenon that occurs due to the interaction of charged particles of the solar wind with the Earth's magnetic field, most often visible in polar latitudes. In the Northern Hemisphere, it is called aurora borealis, and in the Southern Hemisphere, it is called aurora australis. Many other planets in the Solar System also have auroras.

basalt—black or grey eruptive igneous rock that cooled quickly on (near) the Earth's surface.

bathymetry—refers to the method of measurement and the map of the ocean floor, i.e., the depth of the ocean concerning the surface of the ocean (undersea topography; see the term *topography*).

bura (bora)—a strong, dry, and cold wind that blows from the land from the north and northeast along the eastern coast of the Adriatic Sea.

Cambrian—the first period in the geological eon Phanerozoic, in the Paleozoic geological era, which lasted from about 542 million years BC to about 488 million years BC. At the beginning of this period, in the first 10 million years, there was an explosion of living species, that is, multicellular animals, that appeared in the fossil record, also known as the Cambrian explosion.

chemical element—a substance consisting of atoms with the same number of protons in the nucleus. The most abundant chemical element in the universe is hydrogen. The most abundant chemical elements in the Earth's crust are oxygen, silicon, and aluminum, and in the mantle, oxygen, magnesium, and silicon. The Earth's core is an iron and nickel alloy with traces of lighter chemical elements: oxygen, silicon, hydrogen, carbon, and sulfur.

comets—celestial bodies of the Solar System of irregular shape and size from about several hundred meters to several tens of kilometers in cross-section, built of ice, rocks, and dust. They travel mainly in elliptical and parabolic paths around the Sun, and when they are closer to the Sun, due to increased temperature, the ice melts and leaves a trail millions of kilometers long. Orbital periods of comets vary from just a few years to more than 1,000 years.

convection—heat transfer within fluids (in the context of geophysics of the solid Earth, in the liquid outer core, but also the solid mantle) caused by the tendency of warmer (less dense) material to rise against gravity and colder (denser) material to descend due to the influence of gravity. Convection in the mantle occurs at a rate of only a few centimeters per year.

core (Earth's)—the central part of the Earth's interior, consisting of a liquid outer shell, a solid inner part, and, most recently discovered, the solid innermost inner core. According to its chemical composition, it is an alloy of iron and nickel with an admixture of lighter chemical elements.

craton—a stable part of the continental lithosphere that has not been recycled through the Earth's mantle by plate tectonics, usually in the interior of a tectonic plate. It is composed of the oldest rocks on Earth and usually has a thick crust and a lithospheric root down to a depth of several hundred kilometers in the mantle.

cross-correlation—a mathematical measure of the similarity of any two time series; in the context of seismology, two seismograms.

crust (Earth's)—the outer shell of the solid Earth, between the surface and the boundary between the crust and the mantle (*Mohorovičić's Discontinuity*). It can be continental or oceanic.

direct problem—the problem of calculating predictions or theoretical data based on physical properties, for example, calculating the propagation time of seismic body (or surface) waves from a known mathematical model of the Earth's internal structure.

earthquake—sudden release of energy and shaking of the ground due to the movement of seismic waves. A tectonic earthquake occurs when rocks break when external tension caused by tectonic forces overcomes their internal strength. In addition to tectonic, earthquakes can also be volcanic as a result of the collapse of volcanic structures or the movement of magma through the Earth's interior, and can also be due to avalanches, landslides, falling meteorites, tidal forces, and other natural phenomena.

earthquake focus (hypocenter)—the point in the Earth's interior where the rocks begin to break during an earthquake.

electromagnetic force—interaction between two electric charges or charged bodies, which by its nature can be attractive (with opposite charges) or repulsive (with charges of the same sign).

epicenter (of an earthquake)—the closest point on the Earth's surface above the earthquake hypocenter.

erosion—the process of slow movement and removal of solid substances on the surface of the Earth or the surfaces of other planets due to wind, water, or gravitational forces.

fault—planar surface or discontinuity in the rocks along which movement occurs.

force—a physical vector quantity that describes the influence that can change the motion of a body (speed and direction), its shape, or its internal structure. Four fundamental forces can be described that act on elementary particles: gravitational, electromagnetic, weak, and strong nuclear forces (see definitions of *vectors, gravitational force*, and *electromagnetic force*).

frequency—physical quantity for frequency of repetition. The measurement unit of frequency is hertz, where 1 hertz means something repeated once per second.

galaxy—a large cluster of gas, cosmic dust, stars, and their planetary systems. Our Milky Way galaxy consists of 100 to 400 billion stars.

geodesy—an applied science that deals with measuring the size and shape of the Earth, the precise determination of position and orientation in space, and the determination of spatial variations in mass and gravitational acceleration.

geodynamo—a physical mechanism according to which electrically conductive, liquid iron, subject to convection in the outer core and rotation of the Earth, creates a magnetic field (geomagnetic field).

geology—a natural science that deals with the study of the physical structure of the Earth and its history recorded in rocks.

geophysics—a natural science that deals with the study of physical properties and processes of the solid Earth, oceans, and atmosphere, and also includes the study of other planets.

geyser—a thermal source on the surface of the Earth or other planets that periodically ejects water and steam. On some other planets and satellites, we also talk about other gases. Some examples are carbon dioxide on Mars and nitrogen on Triton.

Gondwana—a vast continent or supercontinent in the Southern Hemisphere that consisted of several present-day continents, namely South America, Africa, Australia, and Antarctica, as well as the Arabian Peninsula, the Indian subcontinent, and the island of Madagascar. It began to split into smaller parts in the Jurassic geological period of the Mesozoic era, about 180 million years ago. Together with Laurasia in the north, it formed the supercontinent Pangea.

granite—granular intrusive igneous rock that slowly cooled and solidified in the Earth's interior.

gravitational force (gravity)—a force of mutual attraction between particles or bodies with mass.

Hadean—geological eon, in the period from the creation of the Earth 4.6 billion years ago to 4 billion years ago, when the Archean eon began.

helioseismology—a branch of astrophysics that deals with the study of the interior of the Sun using seismological methods and principles.

hodochrones—curves showing the time it takes for seismic waves to travel from the source of the earthquake to the seismograph at different distances. By their nature, they can be theoretical (based on a mathematical model of the Earth's internal structure and an understanding of the physics of the propagation of seismic waves through the Earth's interior) or empirical (based on recorded data).

Homo erectus—an extinct species of the genus *Homo* in the tribe Hominini, in the family Hominidae, in the order of primates, which is found in the fossil record in the period from about 2 million years ago to approximately 145,000 years BC. The name comes from the Latin for "upright man."

***Homo sapiens* (humans)**—the most widespread living species of the genus *Homo* in the tribe Hominini, in the family Hominidae, in the order of primates. Anatomically modern humans appear in the fossil record for the first time around 300,000 to 200,000 years BC. The name comes from the Latin for "sensible or wise man."

hot spots—volcanic areas on the Earth's surface associated with the rise of hot material in the mantle. Some examples of hot spots are Hawaii, Iceland, Eifel, Azores, Réunion, Samoa, and Yellowstone. An alternative hypothesis involves a less solid or anomalously thin Earth's crust in these areas.

hypocenter—the point in the Earth's interior where rock failure begins during an earthquake.

infrasound—sound waves of frequencies below the frequencies audible to the human ear, that is, below 20 hertz.

inner core (Earth's)—the central part of the Earth's core in a solid state, with a radius of about 1,220 kilometers.

interferogram—a visual representation of the deformation of the Earth's surface in an area taken from a satellite at two different times, for example, before and after an earthquake.

inverse problem—the problem of finding a mathematical or physical model from data, for example, the problem of finding a tomogram of the Earth's internal structure from the time of seismic body wave propagation.

isotope—an atom of the same chemical element that differs in the number of neutrons in the nucleus of the atom. Radioactive isotopes are used for radiometric dating (determining the age of rocks). Some methods are uranium-235–lead-207 (^{235}U to ^{207}Pb) and uranium-238–lead-206 (^{235}U to ^{206}Pb), which are used to determine the age of rocks in the case of zircon.

knot—a unit of speed measurement; 1 nautical mile per hour, which is equivalent to a speed of 1.852 kilometers per hour.

lava—hot, molten rock that erupted from a volcano or a vent on the Earth's surface. It also refers to the cooled rock mass created in the same way.

light year—a unit of measurement of large distances in astronomy; the distance traveled by light in a vacuum in 1 Julian year (365.25 days).

lithosphere—the crust and the uppermost part of the Earth's mantle.

liquefaction (soil)—a phenomenon in which the strength of the soil (most often in sandstone with a granular structure) drastically decreases during the passage of seismic waves so that it begins to behave like a liquid.

loess—sedimentary rock with a dusty structure and yellowish color formed by the action of the wind, mainly in the Pleistocene, a geological epoch from about 1.8 million years ago to 10,000 years BC.

magma—hot, molten rock mass located below the Earth's surface. When it is ejected to the surface of the Earth, it is called lava.

magma chamber—a reservoir of magma inside the Earth's crust under the volcano.

magnetic field—space around natural and artificial magnets in which magnetic forces act. In mathematical terms, a vector field describes the magnetic effect on a charge, current, or material moving through it.

magnetosphere—the space around the planet determined by the action of its magnetic field on charged particles. The shape of the Earth's magnetosphere is changed by the movement and strength of the solar wind—electrically charged particles from the Sun. On the side of the Sun, it is pushed due to the solar wind and extends to distances of about 6 to 10 Earth radii, and on the side opposite to the Sun, it has a "tail" that extends millions of kilometers into interplanetary space.

mantle (Earth's)—the thickest, silicate shell of the Earth, between its crust and core, in a solid state, which behaves like a liquid on the geological scale of millions of years.

meteorites—pieces broken off from comets, asteroids, or even smaller bodies, meteoroids, that hit the surface of the Earth or a planet. According to their composition, they can be silicate, iron, or a combination. Meteorites that burn up in the atmosphere due to friction, that is, do not reach the ground, are called meteors. They are essential for determining the age of the Solar System and understanding the process of planet formation.

microseismic noise—vibrations of the Earth caused by natural phenomena, mainly by the interaction of the mass of ocean water and the atmosphere with the solid Earth, which is recorded by a seismograph.

mid-ocean ridge—an underwater mountain range at the divergent edges of tectonic plates formed by their movement due to the rise of hot material from the mantle. The sea floor is formed from magma that emerges at its crest and moves horizontally over millions of years toward subduction zones.

mineral—a collection of atoms, molecules, or ions connected in a crystal lattice, which can be expressed by a chemical formula.

Mohorovičić's Discontinuity—a transition from the Earth's crust to the mantle, characterized by a difference in rock properties to which seismic waves are sensitive. It was named after Andrija Mohorovičić, who published the discontinuity's discovery in 1910.

moment magnitude—a measure of earthquake magnitude based on the seismic moment, which is more directly related to the energy released during an earthquake than other definitions of magnitude. See the definition of *seismic moment*. There is no unit of measurement.

monsoons—seasonal winds that, in the context of Australia, blow between December and March in its subtropical belt from the west and northwest. The word also refers to the rainy or wet season in the northern parts of Australia.

nautical mile—a unit of measurement of distance on the sea or ocean equal to a length of 1,852 meters, or 1.85 kilometers.

nebula—a large mass of interstellar matter with irregular shapes and unclear boundaries, made of cosmic dust and gases.

nutation—a small periodic oscillation of the precessional cone of the Earth (see the definition of *precession*) due to a change in the inclination of the Moon's path. The most pronounced period of nutation is 18.6 years.

oceanic trench—depression on the ocean floor at convergent boundaries where tectonic plates are pulled under each other (subduction zone).

opal—a type of mineral consisting of silicon dioxide deposited in rock crevices in various colors; it occurs in two categories: common opal or precious opal. In addition to Australia, it is found in Africa, Europe, and the Americas, and its existence has also been confirmed on Mars.

Outback—a remote interior of the Australian continent, very sparsely populated, with a lower or higher degree of remoteness, deeply rooted in Australian heritage, history, and folklore.

outer core—liquid shell between the inner core and the mantle in which convection takes place and the *geodynamo* is generated and maintained.

P waves—longitudinal or compressional body waves that occur during an earthquake and move through the Earth's interior. When these waves pass, the ground particles move in the direction of the waves.

planet—a celestial body in orbit around a star, with enough mass and gravity to form a spherical shape, and with an orbit (the path along which a planet moves around the Sun) devoid of other objects of similar size.

planetesimals—the building blocks of planets that were formed due to the collision and gravity of dust and gas particles in solar nebulae, where new stars and planets orbiting around them are formed.

plate tectonics—the theory that the Earth's lithosphere is composed of several plates that move at speeds of several centimeters per year. Boundaries between plates are convergent when they move toward each other, divergent when the plates move apart, and transform when they move next to each other.

precession—in the astronomical context, a continuous change in the orientation of the Earth's rotation axis due to the gravitational action of the Sun and the Moon. Due to precession, the Earth's axis describes the shape of a cone. The period of precession is 25,772 years.

rocks—solid accumulations of one or more minerals. Constitutive parts of the Earth's internal structure, except in the regions where they are in a liquid state. They are divided into igneous, sedimentary, and metamorphic rocks.

rock strength—a physical quantity that describes the internal property of rocks to resist the action of external forces. It is equal to the stress due to external forces that would cause the rock to break. The unit of measurement for strength is the Pascal.

S waves—shear body waves that occur during an earthquake and move through the Earth's interior. During the passage of these waves, ground particles move perpendicular to the direction of the waves.

San Andreas (fault)—a fault that is also a transform boundary between the North American and Pacific tectonic plates along the California coast, about 1,200 kilometers long.

sand fountain (sand geyser)—a geological phenomenon that occurs in sandstones (porous sedimentary rocks) during and after an earthquake. Seismic waves can cause liquefaction, and under increased pressure, sand mixed with water and gases can begin to rise to the surface.

seismic moment—the moment of a force (torque) in the representation of the source of the earthquake with the help of two pairs of orthogonal forces. The unit of measurement of torque is the Newton meter. It is proportional to the shear properties of the rocks, the average displacement along the fault, and the area of the fault surface that was activated during the earthquake. It serves as a basis for calculating the magnitude of the earthquake moment.

seismic tomography—a method of imaging the underground or the Earth's interior using seismic waves generated by earthquakes or explosions. Tomography is an example of an inverse method.

seismic waves—the waves that move through the Earth (body waves) and its surface (surface waves) as a result of earthquakes, volcanic eruptions, magma movements, landslides, explosions and collapses, meteorite impacts, and other natural and artificially induced phenomena.

seismogram—recording of ground motion by a seismometer. These are displacements, velocities, or accelerations of ground particles recorded in time, either analog or digital.

seismograph—a device used for continuous recording of ground movements due to earthquakes, volcanic eruptions, magma movements, landslides, explosions and collapses, meteorite impacts, microseismic noise, and other natural and artificially induced phenomena. Its essential parts are a sensor, also called a seismometer, and a digitizer that converts ground motion into digital data.

seismology—a branch of geophysics or a scientific discipline that deals with the study of earthquakes and the propagation of seismic waves due to earthquakes and other phenomena through the interior and surface of the Earth. In recent times, it also refers to other planets and stars.

seismometer—a seismograph sensor that is sensitive to and records ground motion.

seismoscope—a primitive device for detecting ground motion due to earthquakes.

silicate—the dominant mineral in the rocks of the Earth's crust and mantle, consisting of silicon and oxygen, with admixtures of other chemical elements.

Solar System—the system of our parent star, the Sun, which consists of eight major planets, nine dwarf planets, thousands of more minor planets, and comets held together by gravity. At a distance of about 28,000 light years from the center of the galaxy, it orbits at a speed of about 220 kilometers per second with an orbital period of about 225 to 250 million years.

spectrogram—a visual representation of the spectrum of recorded frequencies, that is, the strength of each frequency as a function of time; in seismology, the frequency spectrum of seismic waves released during an earthquake.

star—a ball of gas in the interior of which nuclear fusion occurs (the joining of atomic nuclei of lighter chemical elements into heavier ones), large enough to emit light and heat.

stress (pressure)—a physical quantity that describes the action of force on the surface, and in the context of tectonics, tectonic forces due to the interaction between tectonic plates. When the tension overcomes the internal strength of the rocks, an earthquake occurs. The unit of pressure is the Pascal.

subduction zone—a convergent zone where one tectonic plate sinks beneath another. Plates subducting into the Earth's mantle either stagnate in the upper mantle or sink over millions of years to the bottom of the mantle. Volcanoes and earthquakes occur in subduction zones.

surface waves—seismic waves that are generated during an earthquake and move along the Earth's surface. Their amplitudes decrease with distance more slowly than seismic body (P and S) waves.

tectonic plates—large moving lithospheric plates on the Earth's surface, the interaction of which creates earthquakes and volcanoes. There are a dozen larger and several dozen smaller tectonic plates.

tomography (computed tomography, CT)—a method of image reconstruction in a plane or space using any type of penetrating wave. In addition to geophysics, it is used in medicine, astrophysics, archaeology, biology, and other sciences.

tomogram—an image created by tomography.

topography—a scientific discipline that deals with the relief of the Earth or other planets, natural and artificial features on the surface. It also refers to a part of a territory or a map description.

triangulation—method of determining location or distance; in seismology, determining the location of an earthquake with the help of three or more seismograms with which we individually measure the distance to the earthquake; in the context of installing ocean-bottom seismographs, a multi-site sonar location method.

Tropic of Capricorn—a geographical parallel at 23 degrees, 27 arc minutes south of the equator, where during the winter solstice, that is, 21 December, the Sun's rays fall vertically.

ultrasound—sound waves of frequencies above the frequencies audible to the human ear (usually above 20,000 hertz).

vector—a quantity that, in addition to magnitude, also has a direction in space. A good example of a vector is wind speed, while temperature is an example of a quantity with no direction in space (scalar).

volcano—a natural vent or fissure on the Earth's surface through which molten, solid, or gaseous materials erupt from its interior.

zircon—a common mineral in the Earth's crust, which, due to its ability to remain unchanged at high temperatures and pressures, is used to determine the age of rocks using the isotope ratio method. The oldest zircons, 4.4 billion years old, were found in the Yilgarn craton in Australia.

BIBLIOGRAPHY

1. Travelers through time

Asimov, I. 1982. Exploring the Earth and the Cosmos: The growth and future of human knowledge. *Crown Publishers*, New York, USA.

Asimov, I. 1977. The Collapsing Universe. *Hitchinson & Co.*, London, UK.

Jeanloz, R., & Wenk, H.-R. 1988. Convection and anisotropy of the inner core. *Geophysical Research Letters* 15, 72–75.

Pachhai, S., Li, M., Rost, S., Dettmer, J., & Tkalčić, H. 2022. Internal structure of ultralow-velocity zones consistent with origin from a basal magma ocean. *Nature Geoscience* 15, 79–84.

Sagan, C. 1980. Cosmos. *Random House*, New York, USA.

Whitehouse, D. 2015. Journey to the Centre of the Earth: The Remarkable Voyage of Scientific Discovery into the Heart of Our World. Weidenfeld & Nicholson, *Orion Publishing Group*, London, UK.

Wilson, J. T. 1963. Evidence from islands on the spreading of ocean floors. *Nature* 197, 536–538.

https://www.britannica.com/science/extrasolar-planet
https://exoplanets.nasa.gov/
https://en.wikipedia.org/wiki/Exoplanet
https://en.wikipedia.org/wiki/Planetesimal
https://www.anu.edu.au/news/all-news/scientists-discover-leftovers-of
-earth's-dramatic-formation
https://eos.org/articles/layered-zone-beneath-coral-sea-suggests-ancient
-magma-ocean
https://www.nasa.gov/
https://global.jaxa.jp/
https://www.jamstec.go.jp/e/
https://www.esa.int/
https://www.nasa.gov/mission_pages/sunearth/multimedia/magnetosphere
.html

https://en.wikipedia.org/wiki/Age_of_Earth
https://en.wikipedia.org/wiki/Plate_tectonics
https://en.wikipedia.org/wiki/Giant-impact_hypothesis

2. Namazu's tail and Matilda's fate

London, J. 1906. The Story of an Eyewitness. *Collier's Magazine.*

https://www.worldhistory.org/Namazu/
https://en.wikipedia.org/wiki/Tokugawa_shogunate
https://www.nps.gov/pore/index.htm
https://en.wikipedia.org/wiki/Andrew_Lawson
https://en.wikipedia.org/wiki/The_Birds_(film)
http://www.sfmuseum.org/hist6/65twain.html
https://www.hinet.bosai.go.jp/
https://www.berkeley.edu/
https://www.llnl.gov/
https://www.usgs.gov
http://www.gfz.hr/sobe-en/seismographs.htm
https://www.vivino.com/wine-news/the-origin-of-zinfandel-and-primitivo
https://en.wikipedia.org/wiki/Zinfandel

3. Moho's and Inge's binoculars

Birch, A. F. 1940. The alpha-gamma transformation of iron at high pressures, and the problem of the earth's magnetism. *American Journal of Science* 238(3), 192–211.

Brush, S. G. 1980. Discovery of the Earth's core. *American Journal of Physics* 48(9), 705–724.

Dziewoński, A. M., & Anderson, D. L. 1981. Preliminary reference Earth model. *Physics of the Earth and Planetary Interiors* 25, 297–356.

Dziewoński, A. M., & Gilbert, F. 1971. Solidity of the inner core of the Earth inferred from normal mode observations. *Nature* 234, 465–466.

Grubišić, V., & Orlić, M. 2007. Early observations of rotor clouds by Andrija Mohorovičić. *Bulletin of American Meteorological Society* 88, 693–700.

Gutenberg, B. 1913. Über die Konstitution des Erdinnern, erschlossen aus Erdbebenbeobachtungen. *Zeitschrift für Physik* 14, 1217–1218.

Herak, D., & Herak, M. 2007. Andrija Mohorovičić (1857–1936)—on the occasion of the 150th anniversary of his birth. *Seismological Research Letters* 78, 671–674.

Herak, D., & Herak, M. 2010. The Kupa Valley (Croatia) Earthquake of 8 October 1909—100 Years Later. *Seismological Research Letters* 81, 30–36.

Ishii, M., & Dziewoński, A. M. 2002. The innermost inner core of the Earth: Evidence for a change in anisotropic behavior at the radius of about 300 km. *Proceedings of the National Academy of Sciences* 99(22), 14026–14030.

Jeffreys, H. 1926. The rigidity of the Earth's central core. *Geophysical Supplements to the Monthly Notices of the Royal Astronomical Society* 1(7), 371–383.

Kennett, B.L.N., Engdahl, E. R., & Buland, R. 1995. Constraints on seismic velocities in the Earth from traveltimes. *Geophysical Journal International* 122, 108–124

Lehmann, I. 1936. P'. *Publications du Bureau Central Séismologique International* A14(3), S.87–115.

Lehmann, I. 1987. Seismology in the days of old. *Eos Transactions, American Geophysical Union* 68(3), 33–35.

Mohorovičić, A. 1910a. Earthquake of 8 October 1909. *Geofizika* 9, 3–55.

Mohorovičić, A. 1910b. Potres od 8. X 1909. Godišnje izvješće Zagrebačkog meteorološkog opservatorija za godinu 1909. 9(4), 1–56.

Mohorovičić, A. 1910c. Das Beben vom 8. X. 1909. Jahrbuch des meteorologischen Observatoriums in Zagreb (Agram) für das Jahr 1909. 9(4), 63 pp.

Oldham, R. D. 1906. The constitution of the interior of the Earth, as revealed by earthquakes. *Quarterly Journal of the Geological Society* 62(1–4), 456–475.

Orlić, M. 2007. Andrija Mohorovičić as a meteorologist. *Geofizika* 24(2), 75–91.

Orlić, M. 2019. How the discontinuity became Mohorovičić's. 31–75. https://doi.org /10.21857/mwo1vcz70y. In Kroz koru do plašta: nove spoznaje o Andriji Mohorovičiću (1857.-1936.), ed. Paušek-Baždar, S., Ilakovac, K., & Orlić, M. *Croatian Academy of Sciences and Arts (HAZU)*, Zagreb, Croatia.

Romanowicz, B., Li, X.-D., & Durek, J. 1996. Anisotropy in the inner core: Could it be due to low-order convection? *Science* 274, 963–966.

Skoko, D., & Mokrović, J. 1982. Andrija Mohorovičić. 1st ed. *Školska knjiga*, Zagreb, Croatia.

Shearer, P. M. 2009. Introduction to Seismology. 2nd ed. *Cambridge University Press*, Cambridge, UK.

Stipčević, J., Tkalčić, H., Herak, M., Markušić, S., & Herak, D. 2011. Lithospheric structure of Croatia from teleseismic receiver functions. *Geophysical Journal International* 185, 1103–1119.

Tkalčić, H. 2016. Andrija Mohorovičić—an extraordinary scientist who pointed his binoculars down. *Seismological Society of Japan, Naifuru* 105, 6–7. http://www .zisin.jp/publications/pdf/nf-vol105.pdf.

Tkalčić, H., & Phạm, T.-S. 2018. Shear properties of the Earth's inner core constrained by a detection of J waves in global correlation wavefield. *Science* 362(6412), 329–332.

Tkalčić, H., Phạm, T.-S., & Wang, S. 2020. The Earth's coda correlation wavefield: Rise of the new paradigm and recent advances. *Earth-Science Reviews* 208, 103285.

Wiechert, E. 1897. Über die Massenverteilung im Inneren der Erde. Nachrichten von der Gesellschaft der Wissenschaften zu Göttingen. *Mathematisch-Physikalische Klasse*, 221–243.

Wegener, A. 1912. Die Entstehung der Kontinente. *Geologische Rundschau* 3(4), 276–292.

https://en.wikipedia.org/wiki/Mohorovičić_discontinuity
https://www.earthscrust.org.au/science/historic/andrija.html

video of the American Geophysical Union, *Icons in Discovery: Inge Lehmann* https://www.youtube.com/watch?v=w2Tj-8FJFeY

https://www.anu.edu.au/news/all-news/anu-researchers-confirm-earth's-inner-core-is-solid
https://www.smh.com.au/education/science-of-earths-core-takes-a-dramatic-twist-20150225-13oz5m.html
https://phys.org/news/2018-10-earth-core-solid.html
https://en.wikipedia.org/wiki/Inge_Lehmann
http://scihi.org/emil-wiechert-inner-structure-earth/
https://en.wikipedia.org/wiki/Chikyū
https://www.globaltimes.cn/page/202401/1305206.shtml
https://www.vesselfinder.com/vessels/details/9947108

4. Giants that sometimes wake up

Shearer, P. M. 2009. Introduction to Seismology. 2nd ed. *Cambridge University Press*, Cambridge, UK.

Sylvander, M., & Mogos, D. G. 2005. The sounds of small earthquakes: Quantitative results from a study of regional macroseismic bulletins. *Bulletin of Seismological Society of America* 95, 1510–1515.

Tosi, P., Sbarra, P., & De Rubeis, V. 2012. Earthquake sound perception. *Geophysical Research Letters* 39(24), L24301.

https://en.wikipedia.org/wiki/Modified_Mercalli_intensity_scale
https://en.wikipedia.org/wiki/Moment_magnitude_scale
https://en.wikipedia.org/wiki/1880_Zagreb_earthquake
https://sh.wikipedia.org/wiki/Panonsko_more
https://en.wikipedia.org/wiki/Pannonian_Sea
https://en.wikipedia.org/wiki/1880_Zagreb_earthquake

5. On the crests of waves

Herak, D., & Herak, M. 2010. The Kupa Valley (Croatia) Earthquake of 8 October 1909—100 Years Later. *Seismological Research Letters* 81, 30–36.

Hutko, A. R., Bahavar, M., Trabant, C., Weekly, R. T., Van Fossen, M., & Ahern, T. 2017. Data products at the IRIS-DMC: Growth and usage. *Seismological Research Letters* 88, 3.

Kennett, B.L.N. 2002. The Seismic Wavefield: Volume 2, Interpretation of Seismograms on Regional and Global Scales. Illustrated ed. *Cambridge University Press*, Cambridge, UK.

Love, A.E.H. 1927. A Treatise on the Mathematical Theory of Elasticity. 4th ed. *Cambridge University Press*, Cambridge, UK.

O'Malley, R. T., Mondal, D., Goldfinger, C., & Behrenfeld, M. J. 2018. Evidence of systematic triggering at teleseismic distances following large earthquakes. *Scientific Reports* 8, 11611.

Prelogović, E., Saftić, B., Kuk, V., Velić, J., Dragaš, M., & Lučić, D. 1998. Tectonic activity in the Croatian part of the Pannonian basin. *Tectonophysics* 297 (1–4), 283–293.

Rayleigh, J.W.S. 1885. On waves propagating along the plane surface of an elastic solid. *Proceedings of the London Mathematical Society* 17, 4–11.

Rösler, B., & van der Lee, S. 2020. Using seismic source parameters to model frequency-dependent surface-wave radiation patterns. *Seismological Research Letters* 91, 992–1002.

Sagan, Carl. 1994. Pale Blue Dot: A Vision of the Human Future in Space. *Random House*, New York, USA

Ulaby, F. T., & Long, D. G. 2014. Microwave radar and radiometric remote sensing. *University of Michigan Press*, Ann Arbor, USA.

Uys, D. 2016. InSAR: An introduction. *Preview* 2016(182), 43–48. https://doi.org/10.1071/PVv2016n182p43

https://en.wikipedia.org/wiki/1667_Dubrovnik_earthquake
https://en.wikipedia.org/wiki/1963_Skopje_earthquake
https://en.wikipedia.org/wiki/1969_Banja_Luka_earthquake
https://en.wikipedia.org/wiki/1979_Montenegro_earthquake
https://en.wikipedia.org/wiki/1996_Ston–Slano_earthquake
https://en.wikipedia.org/wiki/Miranda_(moon)
https://temblor.net/
https://temblor.net/earthquake-insights/petrinja-earthquake-moved-crust-10-feet-12410/

6. A sharp look inside

Bayes, T. 1763. An essay towards solving a problem in the doctrine of chances. *Philosophical Transactions* 53, 370–418.

Belonoshko, A. B., Lukinov, T., Fu, J., Zhao, J., Davis, S., & Simak, I. S. 2017. Stabilization of body-centered cubic iron under inner-core conditions. *Nature Geoscience* 10, 312–316.

Cormier, V. F., Bergman, M. I., & Olson, P. L. 2022. Earth's Core: Geophysics of a Planet's Deepest Interior. *Elsevier*, Cambridge, MA, USA.

Dziewoński, A. M., & Anderson, D. L. 1981. Preliminary reference Earth model. *Physics of the Earth and Planetary Interiors* 25, 297–356.

Ishii, M., & Dziewoński, A. M. 2002. The innermost inner core of the Earth: Evidence for a change in anisotropic behavior at the radius of about 300 km. *Proceedings of the National Academy of Sciences* 99(22), 14026–14030.

Kennett, B.L.N., Engdahl, E. R., & Buland, R. 1995. Constraints on seismic velocities in the Earth from traveltimes. *Geophysical Journal International* 122, 108–124.

Li, Y., Miller, M. S., Tkalčić, H., & Sambridge, M. 2021. Small-scale heterogeneity in the lowermost mantle beneath Alaska and northern Pacific revealed from shear-wave triplications. *Earth and Planetary Science Letters* 559, 116768.

McDonough, W. F., & Sun, S. S. 1995. The composition of the Earth. *Chemical Geology* 120, 223–253.

Mousavi, S., Tkalčić, H., Hawkins, R., & Sambridge, M. 2021. Lowermost mantle shear-velocity structure from hierarchical trans-dimensional Bayesian tomography. *Journal of Geophysical Research* 126(9), e2020JB021557.

Muir, J., Tanaka, S., & Tkalčić, H. 2022. Long-wavelength topography and multiscale velocity heterogeneities at the core-mantle boundary. *Geophysical Research Letters* 49(7), e2022GL099943.

Pachhai, S., Li, M., Rost, S., Dettmer, J., & Tkalčić, H. 2022. Internal structure of ultralow-velocity zones consistent with origin from a basal magma ocean. *Nature Geoscience* 15, 79–84.

Poupinet, G., Pillet, R., & Souriau, A. 1983. Possible heterogeneity of the Earth's core deduced from PKIKP travel times. *Nature* 305, 204–206.

Romanowicz, B. 2003. Global mantle tomography: Progress status in the last 10 years. *Reviews of Geophysics and Space Physics* 31(1), 303.

Romanowicz, B. 2008. Using seismic waves to image Earth's internal structure. *Nature* 451, 266–268.

Sambridge, M., Bodin, T., Gallagher, K., & Tkalčić, H. 2013. Transdimensional inference in the geosciences. *Philosophical Transactions of the Royal Society A* 371, 20110547.

Stephenson, J., Tkalčić, H., & Sambridge, M. 2020. Evidence for the innermost inner core: Robust parameter search for radially varying anisotropy using the Neighbourhood Algorithm. *Journal of Geophysical Research* 126, e2020JB020545.

Tkalčić, H. 2017. *The Earth's Inner Core: Revealed by Observational Seismology.* Cambridge University Press, Cambridge, UK.

Tkalčić, H. 2015. Complex inner core of the Earth: The last frontier of global seismology. *Reviews of Geophysics* 53(1), 59–94.

Tkalčić, H., & Phạm, T.-S. 2018. Shear properties of the Earth's inner core constrained by a detection of J waves in global correlation wavefield. *Science* 362(6412), 329–332.

Williams, Q., & Garnero, E. J. 1996. Seismic evidence for a partial melt at the base of the mantle. *Science* 273, 1528–1530.

https://en.wikipedia.org/wiki/Kola_Superdeep_Borehole

https://www.bbc.com/future/article/20190503-the-deepest-hole-we-have
-ever-dug

https://www.space.com/15139-northern-lights-auroras-earth-facts-sdcmp.html

https://www.nationalgeographic.org/encyclopedia/aurora/

https://en.wikipedia.org/wiki/The_Core

https://en.wikipedia.org/wiki/Tomography

https://www.nibib.nih.gov/science-education/science-topics/computed
-tomography-ct

https://mathworld.wolfram.com/VoronoiDiagram.html

https://en.wikipedia.org/wiki/Voronoi_diagram

https://cosmosmagazine.com/earth/earth-sciences/most-detailed-look-yet-at
-ultra-low-velocity-zones/

https://eos.org/articles/layered-zone-beneath-coral-sea-suggests-ancient
-magma-ocean

https://www.anu.edu.au/news/all-news/scientists-discover-leftovers-of
-earth's-dramatic-formation

https://www.iflscience.com/physics/mysterious-structures-near-earths-core
-could-be-legacy-of-moonforming-impact/

http://www.sci-news.com/othersciences/geoscience/earths-innermost-inner
-core-09425.html

https://www.anu.edu.au/news/all-news/scientists-dig-deep-to-reveal-earth's
-hidden-layer

https://www.science.org/content/article/scientists-probing-secrets-earths
-inner-core-saved-life-planet

https://www.abc.net.au/news/science/2022-02-10/earth-inner-core
-superionic-iron-element-solid-liquid-geophysics/100813000

7. Dragon's jaw and crystal ball

Bolt, B. 1978. Earthquakes: A Primer. *W.H. Freeman*, San Francisco, USA, ISBN 0-7167-0094-8.

Hough, S. E., 2002. Earthshaking Science. *Princeton University Press*, Princeton, NJ, USA, ISBN 9780691118192.

Jones, L. 2018. The Big Ones: How natural disasters have shaped us (and what we can do about them). *Icon Books*, London, UK, ISBN 9781785784361.

Wang, K., Cheng, Q., Sun, S., & Wang, A. 2006. Predicting the 1975 Haicheng Earthquake. *Bulletin of Seismological Society of America* 96(3), 757–795.

https://en.wikipedia.org/wiki/Zhang_Yimou
https://en.wikipedia.org/wiki/Aftershock_(2010_film)
https://en.wikipedia.org/wiki/Sophie's_Choice_(film)
http://stephaneonblogger.blogspot.com/2016/01/the-deadliest-earthquake-in-history-was.html
https://www.sciencefriday.com/articles/earthquake-prediction/
https://www.scientificamerican.com/article/italian-scientists-get/
https://www.science.org/content/article/italy-s-supreme-court-clears-l-aquila-earthquake-scientists-good
https://www.nature.com/articles/515171a
http://www.news.cn/english/2021-11/22/c_1310325922.htm

8. Red

Alvarez, W. 1997. *T. rex* and the Crater of Doom. *Princeton University Press*, Princeton, NJ, USA.

Debayle, E., Kennett, B.L.N., & Priestley, K. 2005. Global azimuthal seismic anisotropy: the unique plate-motion deformation of Australia. *Nature* 433, 509–512.

Goldstein, P., & Snoke, A. 2005. SAC availability for the IRIS community. *Incorporated Institutions for Seismology Data Management Center Electronic Newsletter.* http://www.iris.edu/news/newsletter/vol7no1/page1.htm

Kennett, B.L.N., & van der Hilst, R. D. 1996. Using a synthetic continental array in Australia to study the Earth's interior. *Journal of Physics of the Earth* 44, 669–674.

Tkalčić, H., Flanagan, M., & Cormier, V. 2006. Observations of near-podal P'P' precursors: Evidence for back scattering from the 150–220 km zone in Earth's upper mantle. *Geophysical Research Letters* 33, L03305.

Tkalčić, H. 2017. The Earth's Inner Core: Revealed by Observational Seismology. *Cambridge University Press*, Cambridge, UK, ISBN 9781107037304.

van der Hilst, R. D., Kennett, B.L.N., & Shibutani, T. 1998. Upper mantle structure beneath Australia from portable array deployments, 39–58. In *Structure and*

Evolution of the Australian Continent, ed. J. Braun et al., *American Geophysical Union Geodynamics Series* 26.

Wessel, P., Smith, W.H.F., Scharroo, R., Luis, J., & Wobbe, F. 2013. Generic mapping tools: Improved version released. *EOS, Transactions American Geophysical Union* 94(45), 409–410. https://doi.org/10.1002/2013EO450001

Worsley, F. A. 1931. Endurance: An Epic of Polar Adventure. *W. W. Norton*, New York, USA, 1999.

https://earthsciences.anu.edu.au/research/facilities/warramunga-seismic-and-infrasound-research-station

https://www.australian-trains.com/ghan

https://ds.iris.edu/ds/nodes/dmc/

https://www.iris.edu/hq/files/publications/annual_reports/doc/1998_AR.pdf

https://seismo.berkeley.edu/history/

https://en.wikipedia.org/wiki/Northern_Exposure

http://rses.anu.edu.au/~nick/wombat/

https://en.wikipedia.org/wiki/Lichfield

https://www.ctbto.org/

https://en.wikipedia.org/wiki/Treaty_on_the_Non-Proliferation_of_Nuclear_Weapons

9. At the bottom of the ocean

Cande, S. C., & Stock, J. M. 2004. Pacific–Antarctic–Australia motion and the formation of the Macquarie Plate. *Geophysical Journal International* 157(1), 399–414.

Coffin, M., Fann, Y., Stock, J. M., Tkalčić, H., Eakin, C., & Macquarie Ridge 3-D Team. 2021. Potentially tsunamigenic mass wasting along Macquarie Ridge, the Australia-Pacific plate boundary, southwest Pacific Ocean. *NH54A01, American Geophysical Union Fall Meeting*, New Orleans, LA, USA.

Meckel, T. A., Coffin, M. F., Mosher, S., Symonds, P., Bernardel, G., & Mann, P. 2003. Underthrusting at the Hjort Trench, Australian-Pacific plate boundary: Incipient subduction? *Geochemistry, Geophysics, Geosystems* 4(12), 1099.

Tkalčić, H., Eakin, C., Coffin, M. F., Rawlinson, N., & Stock, J. M. 2021a. Deploying a submarine seismic observatory in the furious fifties. *Eos Transactions, American Geophysical Union*, 102.

Tkalčić, H., Eakin, C., Rawlinson, N., Coffin, M., Stock, J. M., & Macquarie Ridge 3D Team. 2021b. Probing the Australia-Pacific Plate Boundary: Macquarie Ridge in 3-D. *S21B-01, American Geophysical Union Fall Meeting*, New Orleans, LA, USA.

Tkalčić, H., Eakin, C., Rawlinson, N., Coffin, M. F., and Stock, J. M. 2020. *Macquarie Ridge* [Data set]. AusPass: The Australian Passive Seismic Server. https://doi.org /10.7914/SN/3F_2020.

Young, M., Rawlinson, N., Arroucau, P., Reading, A. M., & Tkalčić, H. 2011. High frequency ambient noise tomography of southeast Australia: New constraints on Tasmania's tectonic past. *Geophysical Research Letters* 38, L13313.

https://mnf.csiro.au/
https://www.csiro.au/en/
https://www.arc.gov.au/
https://www.antarctica.gov.au/
https://www.nerc.com/Pages/default.aspx
https://www.antarctica.gov.au/antarctic-operations/stations/macquarie-island/
https://www.antarctica.gov.au/news/2021/shaking-up-earthquake-science-on -macquarie-island/
https://en.wikipedia.org/wiki/Macquarie_Island

10. New worlds and their explorers

Dreger, D., Tkalčić, H., & Johnston, M. 2000. Dilatational processes accompanying earthquakes in the Long Valley Caldera. *Science* 288, 122–125.

Garcia, R. F., Gagnepain-Beyneix, J., Chevrot, S., & Lognonné, P. 2011. Very preliminary reference Moon model. *Physics of the Earth and Planetary Interiors* 188, 96–113.

Head, J. W., III & Coffin, M. F. 1997. Large igneous provinces: A planetary perspective. *American Geophysical Union Geophysical Monograph* 100, 411–438.

Khan, A., et al. 2021. Upper mantle structure of Mars from InSight seismic data. *Science* 373, 434–438.

Knapmeyer-Endrun, B., et al. 2021. Thickness and structure of the Martian crust from InSight seismic data. *Science* 373, 438–443.

Lognonné, P., et al. 2020. Constraints on the shallow elastic and anelastic structure of Mars from InSight seismic data. *Nature Geoscience* 13, 213–220.

Phạm, T.-S., & Tkalčić, H. 2021. Constraining floating ice shelf structures by spectral autocorrelation of teleseismic P-wave coda: Ross Ice Shelf, Antarctica. *Journal of Geophysical Research* 126(4), e2020JB021082.

Romanowicz, B., Li, X.-D., & Durek, J. 1996. Anisotropy in the inner core: Could it be due to low-order convection? *Science* 274, 963–966.

Sambridge, M., Tkalčić, H., & Jackson, A. 2010. Benford's law in the natural sciences. *Geophysical Research Letters* 37, L14312.

Stähler, S. C., et al. 2021. Seismic detection of the Martian core. *Science* 373, 443–448.

Sun, W., & Tkalčić, H. 2022. Repetitive marsquakes in Martian upper mantle. *Nature Communications* 13, 1695.

Tkalčić, H., & Sun, W., 2023. Marsquakes redefine what we thought about quiet Mars. *Science Breaker.* https://doi.org/10.25250/thescbr.brk676.

Wang, S., & Tkalčić, H. 2022. Scanning for planetary cores from inter-source correlations via a single seismograph. *Nature Astronomy* 6, 1272–1279.

Weber, R. C., Lin, P., Garnero, E. J., Williams, Q., & Lognonné, P. 2011. Seismic detection of the lunar core. *Science* 331, 309–312.

Wieczorek, M., et al. 2006. The constitution and structure of the lunar interior. *Reviews in Minerology and Geochemistry* 60, 221–364.

Yokoo, S., Hirose, K., Tagawa, S., Morard, G., & Ohishi, Y. 2022. Stratification in planetary cores by liquid immiscibility in Fe-S-H. *Nature Communications* 13, 644.

https://mars.nasa.gov/insight/mission/overview/
https://mars.nasa.gov/insight/
https://honors.agu.org/fellows/fellows-alpha-list/
https://en.wikipedia.org/wiki/Apollo_program
https://www.nasa.gov/mission_pages/apollo/missions/index.html
https://mars.nasa.gov/mars-exploration/missions/viking-1-2/
https://www.seis-insight.eu/en/seis-news/517-seis-results
https://phys.org/news/2016-08-earthquakes-oceans.html
https://www.jpl.nasa.gov/
https://mathworld.wolfram.com/BenfordsLaw.html
https://www.nasa.gov/content/solar-missions-list
https://www.fleetspace.com/newsroom/fleet-space-awarded-4m-demonstrator-program-contract-to-harness-seismic-data-at-moons-south-pole

INDEX

Aboriginal, 9, 158n2, 178, 195

Aconcagua, Argentina, 211

Adelaide, Australia, 159, 171

Adria, tectonic microplate, 110

Adriatic, the cost of, 34, 37, 71, 97;
Sea, 37, 79, 89, 111; microplate of,
88. *See also* Adria

Afghanistan, 159n, 167

Africa, 21, 127, 129, 192; caves in, 151;
North Africa, 191; subduction
under, 21; tectonic plate of 110

aftershocks, 23, 43, 71–72, 74–75,
79–80, 83, 85, 107, 149, 167;
distribution of, 167; film *Aftershock*,
138–39, 139n2, 151; frequency of,
7, 114; locations and times of, 76.
See also main shock

Albania, 70

Aldrin, Buzz, 10, 161, 224, 226

Aleutian Islands, 190

Alice Springs, Australia, 158,
185–87

Alvarez, Louis, 183n

Alvarez, Walter, 183n

America, vii; tectonic plate of, 21, 31,
36, 153; coast of, 19, 42, 74, 190;
continent of, 2, 19, 29, 31

American Geophysical Union, 61n3,
149n, 214n, 224n

Americans, 228; and Civil War, 175;
nuclear tests of, 183; observatories
and universities of, 33; scientists of,
42n, 61, 224n, 248n1; and
Revolution, 176

Amiel, Jon, 119n

Anatolia, 38. *See also* Anatolian
Fault; Turkey

Anatolian Fault, East, 91; North, 38;
system of, 91

Anderson, Don, 61n1, 227

anisotropy, 133–134; of inner
core, 228n

Antarctic Circumpolar Current, 9,
200, 203, 206, 214–15

Antarctica, 9, 163n1, 167n, 199, 204,
232; tectonic plate of, 204

Apennine peninsula, 38, 42–43

Apollo missions, 10, 225–26

Arafura Sea, 172

Arcidiaco, Armando, 173

Argentina, 191

Aristotle, 87

Armstrong, Neil, 10, 161, 224, 226

Asia, vii, 140–41, 191; language
of, 239

asteroids, 16, 20; impact of, 183, 183n

astronomers, 33, 225n; interest of,
161; knowledge of, 13

radar, 102, 209
radiation (pattern) diagrams, 98–100
Rawlinson, Nick, 165, 196, 207n1,
   212
Rayleigh, Lord (John William Strutt),
   95. *See also* Rayleigh waves
Rayleigh waves, 95–96, 99–101.
   See also surface waves
Reading, Anya, 167n
Reid, Harry F., 35
Research School of Earth Sciences
   (RSES), 164n, 179
Reuber-Paschwicz, Ernst von, 53n1
Reynolds River, Australia, 175n
Rhea, 87
Rhie, Junkee, 160n
Richards, Mark, 31
Richter, Charles, 65–68; *See also*
   Richter magnitude; Gutenberg-
   Richter law
Richter magnitude, 2, 67
Ridgecrest, California: earthquake of
   2019, 89; earthquake of 2020, 89
Rijeka, Croatia, 39, 94
Ring of Fire, 141, 192
Romanowicz, Barbara, 228, 228n
Rome: emperor of, 58; foundations
   of, 58
RV Investigator, 188–89, 193, 196,
   198–99, 207, 218–19, 223
Ryugu (asteroid), 16

S-waves (shear waves). *See* seismic
   waves
Sagan, Carl, 224, 224n, 237
Sahara, 178
Sambridge, Malcolm, 131n2, 248n2
San Andreas Fault, 5, 19, 31, 33–36,
   43, 73–74, 91, 152–53

San Francisco (San Francisco Bay
   Area), California, 2, 30, 32–34, 152,
   161, 228; agony of, 32; and
   earthquakes, 33; population of, 33;
   See also San Francisco, California,
   earthquake of 1906
San Francisco, California, earthquake
   of 1906 (the Great San Francisco
   Earthquake and Fire), 5, n28, 30,
   32, 42–44, 112, 116, 150, 162
San Juan Bautista, California, 34
Santiago de Chile, Chile, 159
Saturn, 253; satellites of, 253n2
Sea of Tranquility, The Moon, 161
SEIS, 230, 238–42, 244–45. *See also*
   InSight
seismic waves, 2–3, 6–7, 18, 24–25,
   30, 36, 40, 40n2, 44, 48–50, 53–56,
   89, 110–11, 115, 119–24, 127n,
   133–135, 157, 179, 190, 204, 209,
   212; crests and troughs of, 6; and
   inner core, 59–62, 161–63, 190–92;
   and meteorites, 238; on Mars,
   239–41, 244–49; and nuclear
   explosions, 163, 181–82; physics of,
   53; propagation of, 44–62, 92–101,
   123, 128, 150, 153, 155, 161–163,
   208, 227, 231; scattering of, 245;
   speed (velocity) of, 54, 60, 124;
   and sound, 82–85; in stars, 252,
   252n; types of waves, 18, 44,
   48–62, 58, 92–101, volumetric
   coverage of, 124–25, 231. *See also*
   surface waves
seismology, vii,2, 5–6, 27, 30n1,
   56–57, 61n3, 71, 133; and
   astronomy, 65; blue-planet
   (environmental), 212, 212n; and
   Earth structure, 39, 133, 235;

Uhrhammer, Bob, 162
Uluru (Ayers Rock), Australia, 169–71
ultra-low velocity zones (ULVZ), 130–33, 130n2
United States Geological Survey (USGS), 35, 35n, 43, 88, 106, 154, 154n
United States of America, 33, 138n, 181, 191, 225, 228
University of Alaska, Fairbanks, 162
University of California, Berkeley (Berkeley), vii, 29n3, 109n, 161n. *See also* Berkeley
University of Cambridge, 165n, 193, 207n1
University of Connecticut, 90
University of Copenhagen, 59
University of Tasmania, 167n, 193, 207n1
University of Tokyo, 27
University of Zagreb, vii, 228
Uranus, 253; satellite of, 109. *See also,* Miranda

Vallebona, Alessandro, 120
Vanuatu, 125
Venera missions, 226–27
Venus, 21, 21n, 226–27, 229
Verne, Jules, 37–38
Vernić, Silvija, 41
Vernona Rupes, Miranda, 109. *See also* Miranda
Veternica cave, Croatia, 70n
Vienna, Austria: and Mohorovičić, 47–48; and United Nations, 181, 184, 184n
Vietnam, 209

Viking missions, 10, 227, 230, 239; seismographs of, 238, 242–43; probes of, 227, 251
Vinkovci, Croatia, 1, 58, 94
volcanic activity, 13, 21, 39, 104, 133, 183, 229; on Mars, 233, 246, 249–50
volcanic earthquakes, 179, 229, 229n. See also *earthquakes*
volcanoes, 24, 83, 102, 104, 133, 157, 237; extinct, 249; underwater, 231
Volosko, 37, 58, 94
Voronoi, Georgy, 126n3. *See also* Voronoi cells
Voronoi cells, 126
Voyager mission, 109, 117

Wang, Sheng, 209n, 222
Warramunga Seismic and Infrasound Facility (Warramunga Array), 9, 158, 173, 179n, 180–82, 184–86, 184n, 185n, 211
Waramungu (people), 158n2
Waszek, Lauren, 192
Weddell Sea, 163n1
Wegener, Alfred, 48n2
Wen, Hua, 147
Western Australia, 166–68
Wiechert, Emil, 48n2, 53n1
Wiechert seismograph, 27n, 28

X-rays, 120–22, 127
Xingtai, China, earthquake of 1966, 142

Yellowknife, Canada, 182
Yokohama, Japan, 2, 25
Yucatán Peninsula, Mexico, 183n